BLACK CARBON IN THE ENVIRONMENT

BLACK CARBON IN THE ENVIRONMENT

Properties and Distribution

EDWARD D. GOLDBERG

Scripps Institution of Oceanography
University of California at San Diego
La Jolla, California

A WILEY-INTERSCIENCE PUBLICATION

JOHN WILEY & SONS

New York Chichester Brisbane Toronto Singapore

Library of Congress Cataloging in Publication Data:

Goldberg, Edward D.
Black carbon in the environment.

(Environmental science and technology, ISSN 0194-0287)
"A Wiley-Interscience publication."
Includes index.
1. Soot—Environmental aspects. 2. Environmental chemistry. I. Title. II. Series.
TD196.S56G65 1985 628.5 84-27146
ISBN 0-471-81979-4

Printed in the United States of America

10 9 8 7 6 5 4 3 2 1

SERIES PREFACE

Environmental Science and Technology

The Environmental Science and Technology Series of Monographs, Textbooks, and Advances is devoted to the study of the quality of the environment and to the technology of its conservation. Environmental science therefore relates to the chemical, physical, and biological changes in the environment through contamination or modification, to the physical nature and biological behavior of air, water, soil, food, and waste as they are affected by man's agricultural, industrial, and social activities, and to the application of science and technology to the control and improvement of environmental quality.

The deterioration of environmental quality, which began when man first collected into villages and utilized fire, has existed as a serious problem under the ever-increasing impacts of exponentially increasing population and of industrializing society. Environmental contamination of air, water, soil, and food has become a threat to the continued existence of many plant and animal communities of the ecosystem and may ultimately threaten the very survival of the human race.

It seems clear that if we are to preserve for future generations some semblance of the biological order of the world of the past and hope to improve on the deteriorating standards of urban health, environmental science and technology must quickly come to play a dominant role in designing our social and industrial structure for tomorrow. Scientifically rigorous criteria of environmental quality must be developed. Based in part on these criteria, realistic standards must be established, and our technological progress must be tailored to meet them. It is obvious that civilization will continue to require increasing amounts of fuel, transportation, industrial chemicals, fertilizers, pesticides, and countless other products, and that it will continue to produce waste products of all descriptions. What is urgently needed is a total systems approach to modern civilization through which the pooled talents of scientists and engineers, in cooperation with social scientists and the medical profession, can be focused on the development of order and equilibrium in the presently disparate segments of the human environment. Most of the skills and tools that are needed are already in existence. We surely have a right to hope a technology that has created such manifold environmental problems is also capable of solving them. It is our hope that this Series in Environmental Sciences and Technology will not only serve to make this challenge more explicit to the established professionals, but that it also will help to stimulate the student toward the career opportunities in this vital area.

Robert L. Metcalf
Werner Stumm

PREFACE

Black carbon* (often called charcoal, soot, elemental carbon, etc.) is one of the ubiquitous materials circulating around the surface of the earth. It is found in air, soils, sediments, crustal rocks, meteorites, waters, and ices. Its universality is related to its refractory nature with respect to reactions with its surroundings and to its origin in burning processes, which are widespread.

My own interest in black carbon was stimulated by two chance encounters with the material while pursuing environmental problems. The first involved its presence as a nuisance in nets while collecting atmospheric particulates during a cruise across the Atlantic Ocean. The flocs of carbon—called "cokey balls" by Dr. David Parkin of Bath University, a fellow dust collector—darkened the nets and were considered an undesirable addition to our collections. We attributed their origins to incomplete combustion of fossil fuels in the engines of ships.

But more decisive was a visit to the laboratory of Professor Jacques Jedwab of the Free University of Brussels, Belgium, who showed me a scanning electron micrograph of a carbon fragment recovered from a 10-million-year-old manganese module dredged from the seafloor. Jedwab suggested that the particle probably arose from a forest fire in the Tertiary period. My intuition led me to consider the possibility of abundant black carbon particles in marine sediments from natural combustion processes. Collaboratively with Dwight Smith, then at Hope College, Holland, Michigan, and presently at the University of Denver, we sought out quantitative estimates of black carbon in deep sea deposits. Since that time, our laboratories have made additional studies on black carbon in the environment, each one indicating that this material contains a substantial amount of information about the mode of origin and perhaps about the site

*The term black carbon was introduced by Novakov (1984) as "combustion-produced black particulate carbon having a graphitic microstructure."

of its production. Our first studies were done without extensive surveys of the paleobotanical and air pollution literature. However, we were quickly brought out of our isolation and to the realization of the importance of this material in environmental studies.

Black carbon has had a continuing involvement with scientists. Two hundred years ago Sir Percival Pott noticed that chimney sweeps, exposed to high concentrations of soot, had an unusually high frequency of scrotal cancer. This disease probably stems from contacts with such carcinogens as benzopyrene, which are associated with soots (Reif, 1981). With the increased levels of black carbons in the atmosphere as a consequence of fossil fuel and wood burning, the general public is now exposed to the simultaneously produced carcinogens.

Recently, *N*-nitroso organic compounds have been found in atmospheric particulates from the New York area. These compounds, in concentrations equal to the total polynuclear aromatics, are potentially equal in carcinogenicity and probably are associated with the black carbons produced in burning processes (Kneip et al., 1983). An extended understanding of levels and compositions of the organic constituents in particulates is essential for studies of environmentally induced mortalities and morbidities. Characterizations of the major substances are still limited to a few urban and rural localities (Mueller et al., 1982). Kaden et al. (1979) indicated that the observed mutagenicity of black carbons from a variety of fossil fuel soots could be attributed to the combination of nine polycyclic aromatic hydrocarbons. Two of the compounds, perylene and aceperylene, exhibited more mutagenic activity than benzo(a)pyrene.

Further, atmospheric black carbons may be invoked as catalysts for gaseous reactions and as carriers for pollutants to the human lung. These solids may in part determine the relative "life-damaging" qualities of different atmospheres. The contamination of indoor air in prehistoric time is evident through soot on the ceilings of inhabited caves. The inadequate ventilation of open fires was the cause of the polluted atmospheres (Spengler and Sexton, 1983).

The soiling of the atmosphere by coal burning was also well recognized two centuries ago by Count Rumford, who was residing in England (Thompson, 1870–1873):

> The enormous waste of fossil fuel in London may be estimated by the vast dark cloud which continually hangs over the great metropolis; and frequently overshadows the country far and wide; for the dense cloud is certainly composed almost entirely of unconsumed coal, which, having stolen wings from the innumerable fires of this great city, has escaped by the chimneys, and continues to sail about in the air, till having lost the heat which gave it volatility, it falls in a dry shower of extremely black dust

> to the ground, obscuring the atmosphere in its descent, and frequently changing the brightest day into more than Egyptian darkness.
>
> I never view from a distance, as I come into town, this black cloud which hangs over London, without wishing to be able to compute the immense number of caldrons of coal of which it is composed; for, could that be ascertained, I am persuaded so striking a fact would awaken the curiosity and excite the astonishment of all ranks of inhabitants, and perhaps turn their minds to an object of economy to which they have hitherto paid little attention.

Another stimulation to pursue black carbon studies arose from the great need to know more about the carbon cycle and budget, especially in view of the fact that human activities are generating significant amounts of carbon dioxide and black carbons to the atmosphere. The possibility that increasing amounts of carbon dioxide in air may alter climate in the future requires a predictive capability on carbon dioxide levels in the atmosphere. To make such predictions, it is necessary to describe the present distributions and fluxes and to identify significant sources and sinks. Sedimentary black carbon has been proposed as an important sink in the cycle of the element.

Finally, environmental black carbons may have recorded the history of society's involvement with fire, which O. C. Stewart (1956) describes as the "first great force" employed by humans. Along with stone tools and language, the mastery of fire placed humans apart from other primates, perhaps a million years ago. Stewart points out that mankind clearly was aware of fire, started by such natural events as lightning striking or volcanic lavas engulfing biomass. Further, individuals must have husbanded and transported fire before they were able to produce it at will. The earliest evidence of fire associated with a hominid occupation site comes from Chesowanja, Africa, and is dated at 1.42 million years. Burnt clay fragments are the surviving artifacts associated with combustion (Gowlett et al., 1981). Arguments have been made that the earliest use and control of fire by humans took place in the temperate zones, later to be adopted by tropical peoples. However, burned bones found in Ethiopian settlements occupied during the middle Pleistocene bring this concept into question (Kalb et al., 1984).

The first evidences of anthropogenic fires may come from caves, inasmuch as the presence of charcoals in this environment can hardly be attributed to natural causes. Perhaps there is additional evidence from the charcoals within caves to better understand how society employed its "first great force."

Further studies on environmental black carbons might allow an assessment of the deductions of Stewart (1956) regarding the role of prehistoric societies in modifying the surface of the earth through their use of fire. He submits that

> everywhere that man traveled he made campfires and left them to ignite any and all vegetation in the vicinity.

and

> he probably deliberately started conflagrations which swept over the country ... to rouse or drive game during hunting was the reason most frequently recorded over the world (by historic aborigines) ... fire would also be used in the collection of various insects, like crickets, whenever they were used as food .... Burning of grasslands and forests to improve pasture for game has been widely reported .... Of less general applicability and of more specialized use is the setting of vegetation on fire as an act of war. All the reasons for burning over the landscape ... could have existed generally during all the history of man.

There are other early societal uses of fire. The American Indian used it ceremonially as an act of actual or ritual cleansing or as a spectacular entertainment (Pyne, 1982). Pyne also points out that fire was used to maintain the fertility of the soils by the recycling of unused debris and the consequential return of nutrients to the soil. Fields could be cleared of vermin and debris by burning.

There are evidences for either natural or anthropogenic forest burning in the geological record. Stewart points out that the charcoal and herb pollen contents in the Neolithic peat deposits in Europe were more extensive than in previous times, based upon evidence from the geologic record. Other such examples exist. The evidence from botanical studies is not as decisive inasmuch as the possibility of natural causes for a fire event must be eliminated, a task that is often difficult. Stewart indicates three categories of botanical evidence: (1) the dating of fire scars by tree-ring analyses; (2) changes in succession where fire-resistant plants become more dominant; and (3) the comparison of the potential growth of plants from analyses of soil, moisture, temperature, and so on, with actual plant cover.

Some assistance may be found in examining the type of biomass burned. For example, there are very few recorded fires initiated naturally on plains and prairies. Some scientists argue that there are no indisputable records of grass fires initiated by lightning. But perhaps of more importance are answers to the question as to whether environmental charcoals do have records in chemical and physical properties in morphologies, and in abundances that will allow an evaluation of these deductions of Stewart.

Clearly, in the past decade, there has evolved a wealth of concepts and technologies that might lead to some answers.

A rather bizarre involvement of black carbon in a societal problem recently emerged in considerations of the consequences of nuclear war (Turco et al., 1984; Ehrlich et al., 1984). The detonation of nuclear devices may result in the initiation of large-scale forest fires that will inject large amounts of black carbon into the atmosphere. These particles, in addition to the fine crustal dusts raised by the explosions, may reduce the average light levels reaching the earth's surface to such an extent that land temperatures can be lowered to −15° to −25°C—the so-called nuclear winter. Such temperatures can persist for months following explosions of 100 mtons or more and can jeopardize the health of humans and other species that survive the event.

The information in this book is derived from a wide spectrum of scientific pursuits: anthropology, ecology, forestry, atmospheric pollution, chemical engineering, high temperature chemistry, and low temperature geochemistry, among others. It is prepared for environmental scientists as a springboard to an understanding of the disposition, of the stubborn persistance, and of the unique properties of this most unusual substance.

There is a vast literature on black carbons. This volume is derived from a sampling of this literature to illustrate the properties and environmental occurrences of black carbon and of associated materials produced in combustion processes. Cited publications were chosen for their relevance in making a point. Others were omitted for a goal of brevity and, as a consequence, I suspect that some substantial contributions were passed over.

The volume is dedicated to my associates who have pursued additional knowledge about the environmental behaviors and occurrences of black carbon—Dwight Smith, John Griffin, Kathe Bertine, Min Koide, James Herring, and Daniel Suman. It pays tribute to our critics, who, through the peer review process, have several times recommended against support of our work (NSF and EPA), have once cut off our funds in midstream (EPA), have once reduced requested support by a factor of 2 (NSF), and have twice adequately supported our research needs (NSF and DOE). I appreciate the graciousness of Dr. C. S. Wong of the Institute of Ocean Science, Sidney, Victoria, Canada, who provided me with office facilities for a sabbatical stay during which time this volume was initiated.

EDWARD D. GOLDBERG

La Jolla, California
April 1985

CONTENTS

BLACK CARBON IN THE ENVIRONMENT

1

WHAT IS BLACK CARBON?

The black, relatively inert, and ubiquitous forms of carbon found in the environment are referred to as charcoal, fusain, elemental carbon, black carbon, soot, microcrystalline carbon, polymeric carbon, or graphite depending upon the preference of the writer. This impression often leads to confusion in the mind of the reader who may consider the substance as (1) 100% carbon, whether it be graphite, charcoal from wood burning, or some other form; (2) a crystalline mineral with a well-known structure and ability to diffract X-rays and electrons into well-known patterns; or (3) as incompletely burned biomass associated with a complex and varying composition of organic phases.

Environmental scientists often describe this material on the basis of their techniques of isolation or of measurement (see Appendix). One ecologist might define it as the black particles seen under an optical microscope after their chemical separation from a soil or sediment. Another ecologist might include not only the black particles but also the semiburned plant, tree, or grass particles. Geochemists may define the substance on the basis of infrared, Raman, or visible spectroscopy and may quantify its concentrations in environmental samples. Atmospheric scientists might utilize its absorption characteristics in air. On one point they would agree—the material has a highly variable nature.

In this volume, black carbons will encompass those impure forms of the element produced by the incomplete combustion of carbonaceous phases including living and dead biomass as well as fossil fuels. Diamond and metamorphic graphites will be excluded to a large extent from this treatment. Many of the black carbons introduced to the environment come from biomass burning and fall into the category of "charcoals." Guidance to what are black carbons might come from the various descriptions of charcoals. For example, the *McGraw-Hill Encyclopedia of Science and Technology* (1977) provides the following definition:

> Charcoal is a porous solid product containing 85–98% carbon produced by heating carbonaceous materials such as cellulose, wood, peat and coals of bituminous or lower rank at 500–600°C in the absence of air.

The *Encyclopaedia Britannica* (1974) provides additional guidance with:

> Charcoal is an impure form of graphitic carbon obtained as a residue when carbonaceous material is partially burned or heated, with limited access of air. Coke, carbon black and soot may be regarded as forms of charcoal; other forms are often designated by the name of the materials, such as wood, bone, blood, and so on, from which they are derived.

The two definitions are in conflict. The former says charcoal is produced "in the absence of air"; the latter "with limited access of air." The latter alludes to a crystal structure for charcoals by use of the term "graphitic carbon." Yet for some charcoals a graphitic X-ray diffraction pattern is not found.

Black carbon is produced from biomass burning and from fossil-fuel burning. It also may be formed on the earth's surface at temperatures of 200°C or less. Black carbon is impure. Cope (1979) recently has pointed out that environmental charcoals have H/C ratios of 0.25–0.69 and O/C ratios of 0.08–0.33. In addition, they can contain up to a percent or so of nitrogen and sulfur.

Clearly, a more general and a more inclusive definition is needed and one which will satisfy the subject matter covered in this volume. Inasmuch as there are increasing research efforts to study black carbons, a general definition that can be modified in the future as more information becomes available is essential. I propose for the present that black carbon is *an impure form of the element produced by the incomplete combustion of fossil fuels or biomass. It contains over* 60% *carbon with the major accessory elements hydrogen, oxygen, nitrogen, and sulfur.*

The black carbons include the charcoals produced from biomass burning and the soot* from fossil fuel and wood combustion. These two latter terms will also be used throughout this volume.

*Soot is derived from the Old English word "to sit" and probably relates to the products formed by combustion which "sit" on the sides of a chimney. (From *Webster's Third New International Dictionary*, 1965.)

2

THE CHEMICAL AND PHYSICAL PROPERTIES OF BLACK CARBONS

INTRODUCTION

The black carbons formed at high temperatures and found in the environment come from a variety of sources—the combustion of plants, woods, and fossil fuels, and industrially produced substances such as cokes and carbon blacks. The properties of the black carbons, chemical and physical, may be diagnostic of the burned materials and of the combustion itself, as well as the transport processes from the site of their creation.

There are three characteristics of the environmental black carbons that may reveal their sources and of the distance traveled from the source to the site of deposition: their surface morphologies, shapes, and size distributions. These properties have only recently been investigated in some detail. Clearly, there may be additional information within the black carbons, either chemical or physical, that can reveal more about their origins and environmental behaviors.

Although there is an increasing number of chemical and physical studies on naturally produced black carbons, the overwhelming amount of data comes from studies on anthropogenic materials, either of industrial use or from fossil fuel combustion. A detailed review of the chemical and physical properties can be found in Mantell (1968).

Two types of black carbon may be produced from a single combustion source: the smaller, submicron particles that have formed from the vapor phase (see Chapter 3) and the larger particles (tens of microns) that reflect the structure of the burned material or the nature of the burning process, the so-called char. Examples of the latter include the cenospheres from oil burning and the carbonized plant remains which retain some of the original biological structures. Clearly, both types from a single source may have different chemical and physical properties. In the following studies no distinctions between these types have been made, although in some cases only one type would be formed, that is, the burning of acetylene gas to produce a soot that must have formed from a gas phase.

Two characteristics of black carbons probably govern many of their environmental involvements—their chemical inertness at low temperatures (i.e., those temperatures found in waters, airs, and land at the earth's surface) and their surface sorptive properties. At low temperatures black carbon is unreactive with most substances. It does absorb oxygen slowly and releases it upon heating as carbon dioxide.

STRUCTURE

Black carbons may be considered a disordered form of graphite. In the transition from X-ray amorphous substances to graphite, a number of properties change as a consequence of an increase in ordering. This can be seen, for example, in the increase in density going from lampblack to single crystal graphite (Mantell, 1968).

Substance	Density (g/cm^3)
Single crystal graphite	2.25
Carbon, coke	2.05
Carbon, lampblack	1.90

The graphite is composed of platelets of carbon atoms arranged hexagonally. The disorder arises in amorphous carbons from the smaller amount of carbon atoms in the layer planes which have a mean diameter of the order of 25 Å. There are but a small number of layers that lay parallel to each other and those that do are randomly oriented.

Other properties reflect the disorder as well as the presence of impurities, such as heats of oxidation (Mantell, 1968).

Substance	Heat of Oxidation (cal/g)
Graphite to CO_2	7796.6
Amorphous carbon to CO_2	7912
Wood charcoal to CO_2	8080

CHEMICAL COMPOSITIONS

Twenty-nine charcoals from British Mesozoic sediments and some associated coalified phytoclasts were examined for their C, H. O, N, and S contents (Table 2.1) by Cope (1979). The charcoals had a carbon content of 66–87% on a dry weight basis, values similar to those of coals. The coals and charcoals can be distinguished on the basis of their H/C ratios.

Table 2.1. Chemical Compositions of Some Charcoals, Coals, and Chars (Cope, 1979)[a]

	Weight Percent					Atomic Ratios	
Material	C	H	O	N	S	H/C	O/C
Charcoalified phytoclasts	66–87	1.8–4.4	10–30	<0.4–1.6	<0.1–0.4	0.25–0.69	0.1–0.33
Coalified phytoclasts	68–80	4.2–6.0	12–27	<0.7–1.4	0.1–1.0	0.69–0.93	0.1–0.30
Artificial chars	48–66	2.0–6.2	32–45	0.3–0.6	. . .	0.37–1.6	0.4–0.71
Natural chars	50–86	1.8–5.6	11–45	<0.4–0.8	<0.2–0.6	0.27–1.3	0.1–0.67

[a]Reproduced with permission of the authors.

Table 2.2. The Compositions of Some Black Carbons (Medalia and Rivin, 1981)[a]

				Percent Extractable in	
Sample	Percent Elemental Carbon	Percent Ash	Weight Loss (N_2, 910°C)	Methylene Chloride and Toluene	Water
Carbon black (furnace black)	98	0.27	1.5	0.13	0.90
Chimney soot from wood-burning fireplace	0.024	21.8	48.0	15.8	14.2
Blended chimney soots from domestic coal fires	0.36	24.6	52.4	35.6	19.0
Soot from "soot box" of domestic oil furnace	0.83	53.8	43.7	0.64	50.7
Soot from small diesel engine	51	2.2	49.2	51.1	3.6
Urban dust	0.47	64.6	36.2	2.9	27.0

[a]Reproduced with permission of the authors.

The former had a range of 0.69–0.93 while the latter of 0.25–0.69. The oxygen, nitrogen, and sulfur contents are not significantly different for these two groups of carbons. The forest fire and artificially produced charcoals had similar compositions to these phytoclasts, although the H/C and O/C ratios can be significantly higher. Thus, charcoals have a makeup distinctly different from that of pure elemental carbon.

The ash and tar (associated organic molecules) contents of different black carbons appear to be characteristic of their modes of formation (Medalia and Rivin, 1981). The carbon black sample had the highest amount of elemental carbon, whereas coal and oil soots had substantially less (Table 2.2). Chemical analyses (Table 2.3) upon black carbon cenospheres produced in the fluidized bed combustion of coals indicated

Table 2.3. Chemical Analyses of Black Carbon Cenospheres and of the Original Fly Ash (in Parentheses) of a Fluidized Bed Reactor for Coal Combustion (Gay et al., 1984)[a,b]

Element (%)	Coal A		Coal B		Coal C	
Al	1.2	(3.8)	0.7	(3.3)	5.2	(11.9)
Ca	0.5	(0.5)	1.3	(1.5)	1.0	(0.7)
Fe	0.8	(3.7)	0.3	(1.1)	4.9	(6.7)
K	0.1	(0.6)	0.1	(0.4)	1.2	(2.1)
Mg	0.1	(0.1)	0.1	(0.3)	0.2	(0.8)
Si	1.7	(7.6)	1.6	(5.7)	7.9	(19.0)
Ti	0.1	(0.2)	0.1	(0.3)	0.4	(0.3)
S	0.5	(0.4)	0.7	(1.0)	1.0	(0.6)
C	73.5	(37.8)	75.7	(49.1)	45.0	(21.5)
Element (ppm)	**Coal A**		**Coal B**		**Coal C**	
As	10	(27)	7.0	(10.4)	230	(134)
Be	<5	(3.5)	3.2	(2.9)	12.2	(15.2)
Cd	<0.7	(0.1)	0.2	(0.7)	0.3	(0.4)
Co	<19	(<10)	<10	(7.3)	<10	(58)
Cr	40	(692)	21	(76)	222	(292)
Cu	60	(23)	32	(31)	205	(315)
Ni	<70	(48)	47	(65)	372	(227)
Pb	47	(19)	19	(38)	182	(285)
Zn	<40	(60)	36	(86)	279	(326)

[a]Reproduced with permission of the authors and Elsevier Scientific Publishers.
[b]Three coals were analyzed.

lower concentrations of the major elements than found in the fly ash from which they were removed. On the other hand, the trace metals were not markedly depleted in the cenospheres over the fly ash suggesting an enrichment in the black carbons (Gay et al., 1984).

ELECTRON SPIN RESONANCE CHARACTERISTICS

The electron spin resonance (ESR) spectrum of a hexane soot consists of a single symmetrical line devoid of any hyperfine structure (Akhter et al., 1984a). The g value and linewidth are 2.0058 and 3.25 G, respectively, indicating a concordance with literature values of coals and other carbonaceous materials. The hexane soot thus appears to be graphitic in nature and to contain stable free radicals.

Cope (1979) determined the ESR characteristics of some charcoals and coals:

Material	g Value	Linewidth (G)	Free Spin Concentration $\times 10^{18}$ spins/g
Mesozoic charcoals	2.0023–2.0026	0.8–4.1	13–32
Mesozoic coals	2.0022–2.0026	2.4–5.4	5–37
Natural charcoals (recent)	2.0024–2.0032	5.5–7.0	3.1–7.4

The recent charcoals have fewer free spin concentrations and larger linewidths than do the older samples.

ACIDITY

Black carbons may possess acidic or basic characters usually governed by their mode of formation (Chang et al., 1982). Exposure of black carbon to oxygen at temperatures between 200 and 400°C yields an acidic type while treatment at higher temperatures in CO_2 or *in vacuo*, followed by exposure to oxygen, produces a basic form. The ability of carbons to donate protons results from such functional groups as carboxyl and hydroxyl. Chang et al. (op. cit.) indicate that there is still uncertainty regarding the reasons for some black carbons having a basic nature.

CARBON ISOTOPIC COMPOSITION

Carbon-13/carbon-12 ratios were run on the black carbon in atmospheric aerosols collected at the Brookhaven National Laboratory in New York and at Barrow, Alaska. The results in δ C-13, versus the Peedee belemnite (PDB) standard, in parts per thousand were -26.8 and -28.1, respectively (Gaffney et al., 1984). These data are consistent with a predominantly C_3-type plant source and are consistent with the C-14 results, which indicate that over 30% of the black carbon at both sites is due to a biogenic input.

SURFACE FUNCTIONAL GROUPS

A characterization of the functional groups and the extractable compounds in a soot produced by the combustion of hexane has led to a more detailed understanding of the overall structure by Smith and his associates at the University of Denver (Kiefer et al., 1981; Akhter et al., 1984a and 1984b). The soot was produced by the burning of reagent grade *n*-hexane and had a surface area of $89 \pm 2\ m^2/g$. The elemental composition had the following ranges: C, 87–92.5%; H, 1.2–1.6%; and O, 6.0–11%.

The Fourier Transform–Infrared Spectrum displays a complex absorption bands at 1260, 1440, 1590, 1700–1800, and 3040 cm^{-1} (Figure 2.1). A small broad band at 3200 cm^{-1} due to hydrogen-bonded OH also was found. Akhter et al. (op. cit.) indicate the 1590-cm^{-1} band is quite complex resulting from aromatic stretching enhanced by conjugated carbonyls with the carbonyl frequency shifted by a large polyaromatic system to lower wavelengths beneath the aromatic band envelope. Most probably there are contributions to this band from the weak aromatic stretching modes of the graphitic planes, weak at 1585–1580 cm^{-1}.

There are three bands in the 1700–1800 cm^{-1} region: 1720, 1740, and 1775 cm^{-1}. The latter two bands change at the same rate during heating of the black carbon and may respond to a single species. All three disappear at 500°C. The 1720-cm^{-1} band is assigned by Kiefer et al. (1981) to an alkyl carbonyl group, while the other two bands appear to be related to the symmetric and asymmetric stretching modes of a cyclic anhydride. These bands lose half their intensity when the soots are heated to 400–500°C, most probably with the evolution of CO_2.

The 3040- and 1440-cm^{-1} bands occur together and are not observed to form from more oxidizing flames. Akhter et al. (1984a) suggest that

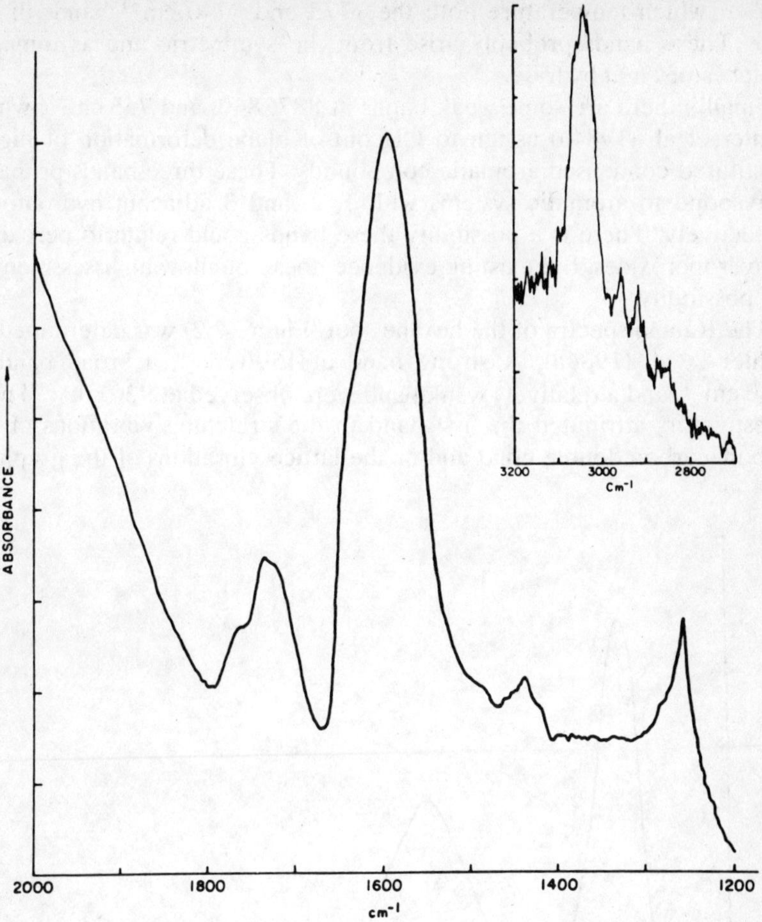

Figure 2.1. Fourier transform infrared spectrum of hexane soot (Kiefer et al., 1981). Reproduced with permission of the authors and the Society of Optical Engineers.

there is strong evidence for these bands relating to unsaturated carbon–hydrogen on the basis of the behavior of the soots following treatment in vacuum at 800°C and reoxidation. Also, the –OH deformation probably contributes to the 1440-cm^{-1} region. These interpretations are in accord with the band shifts when deuterated hexane soots were analyzed.

Akhter et al. (1984a) attribute the band at 1260 cm^{-1} to C–O–C stretching, from both anhydride and aryl ether, as well as a partial contribution from =CH in-plane deformation. The shoulder on the high frequency side of this band disappears when the black carbon is heated to

500° at which temperature both the 1775 and 1740-cm^{-1} bands disappear. These bands probably arise from the symmetric and asymmetric stretches of an anhydride.

Finally, there are some weak bands at 887, 840, and 755 cm^{-1}, which Akhter et al. (1984a) assign to CH out-of-plane deformation of highly substituted condensed aromatic compounds. These three bands probably correspond to aromatic systems with 1, 2, and 3 adjacent hydroatoms, respectively. There is a possibility these bands could relate to peroxides or hydroperoxides, but existing evidence does not allow an assessment of this possibility.

The Raman spectra of the hexane soot (Figure 2.2) was determined by Akhter et al. (1984a). A strong band at 1590 cm^{-1}, a broad band at 1380 cm^{-1}, and a relatively weak band were observed at 920 cm^{-1}. These investigators attributed the first band to the stretching vibrations of the carbon–carbon double bond and/or the lattice vibrations of the graphitic

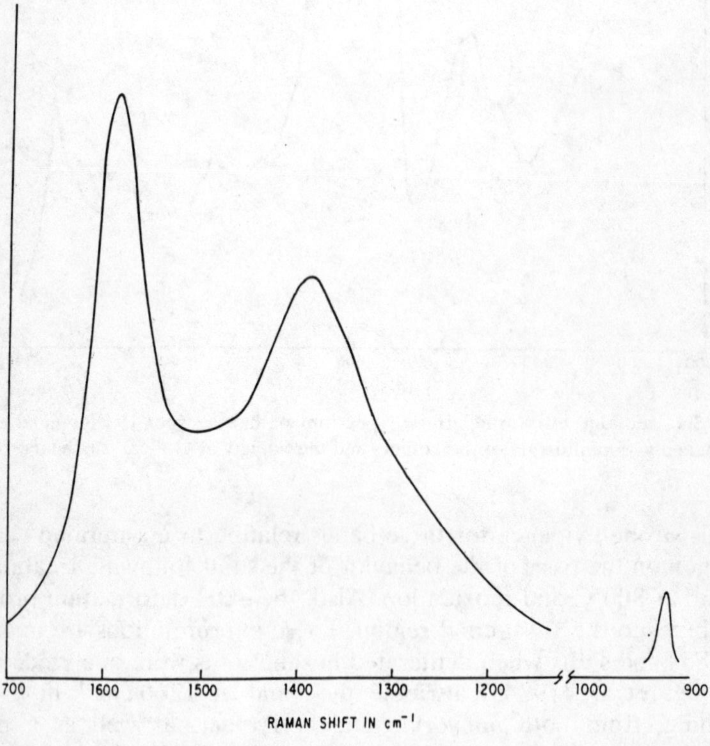

Figure 2.2. Raman spectrum of hexane soot at 23°C. Argon ion laser = 488 nm (Akhter et al., 1984a). Reproduced with permission of the authors and the Optical Society of America.

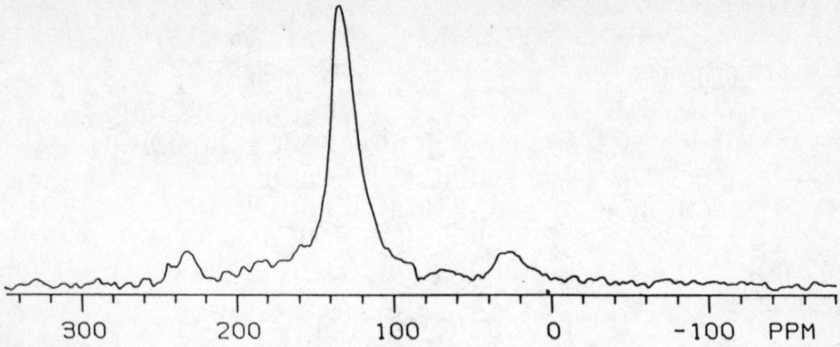

Figure 2.3. CP-MAS C-13 NMR spectrum of hexane soot (Akhter et al., 1984a). Reproduced with permission of the authors and the Optical Society of America.

structure. The 1380-cm^{-1} band probably arises in aromatic skeletal vibrations which are associated with condensed benzene rings. Finally, the 920-cm^{-1} band could be due to the out-of-plane C–H bending vibrations of substituted aromatics or possibly peroxides and/or hydroperoxides.

The aromaticity and the ratios of aromatic to aliphatic carbons were sought by Akhter et al. (1984a) through cross polarization and magic angle spinning C-13 nuclear magnetic resonance (Figure 2.3). The band at 128.175 ppm arises from aromatic and polycyclic aromatic species whereas the aliphatic species are usually found in the region of 11–30 ppm. The aliphatic carbons in the hexane soots are below the limit of detection and thus are assigned a relative concentration of zero. The carbon aromaticity, C_{arom}/C_{org}, is calculated to be 0.89, as a minimum.

On the basis of these spectroscopic studies, a proposed structure for a spherical hexane soot particle (Figure 2.4) has been put forth by Akhter et al. (1984a). The functional groups associated with black carbons include anhydride, a carbonyl conjugated with an aromatic segment, alkyl ketone, aryl ether, and C–H. It appears the relative contributions of these species depend upon the conditions of the black carbon formation.

This proposed structure includes both the skeletal framework of the black carbon as well as any substances included within the pores. Akhter et al. (1984b) through successive extractions of the soluble components of the hexane soot were able to tentatively identify some of the occluded components. The total mass of the extractable compounds was 32.9% of the total and consisted of polyaromatic compounds, oxygenated polyaromatic compounds, and some aliphatic compounds. There was no qualitative change in the infrared spectra going from the original black

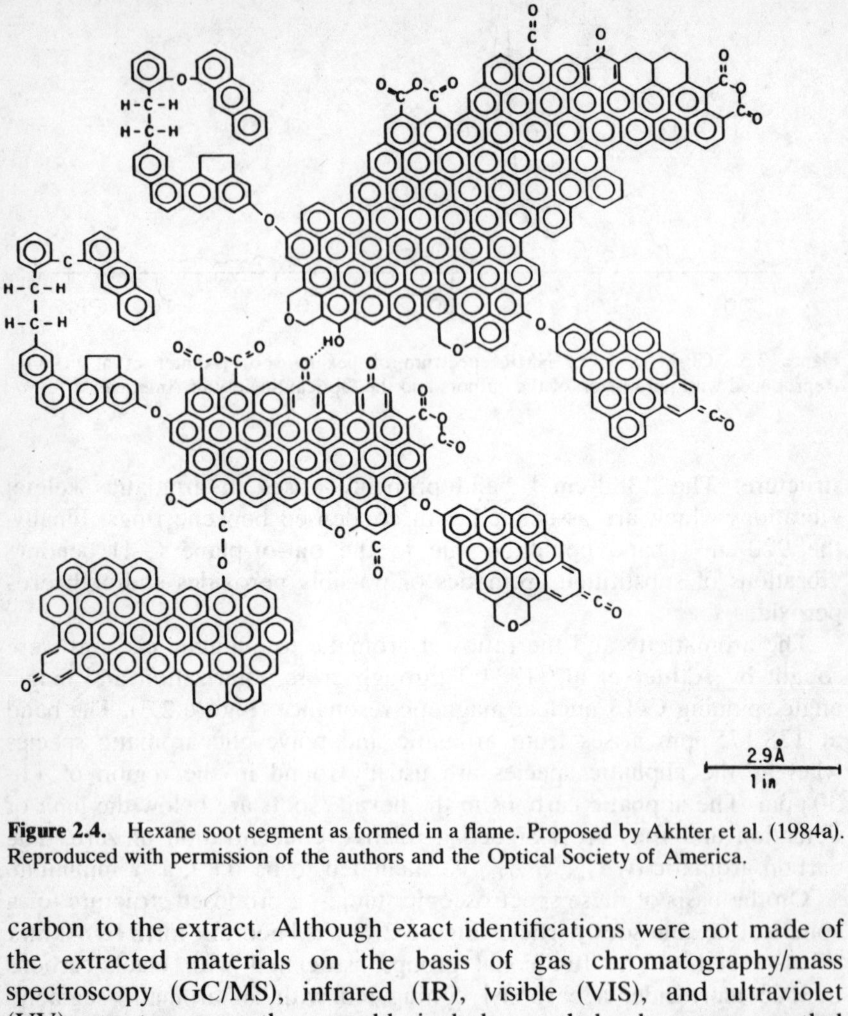

Figure 2.4. Hexane soot segment as formed in a flame. Proposed by Akhter et al. (1984a). Reproduced with permission of the authors and the Optical Society of America.

carbon to the extract. Although exact identifications were not made of the extracted materials on the basis of gas chromatography/mass spectroscopy (GC/MS), infrared (IR), visible (VIS), and ultraviolet (UV) spectroscopy they could include methyl chrysenes, methyl benzo(c)phenanthrene, methyl benz(a)anthracenes, 18-pentatriacontane, dimethyl hexahelicene, triphenyl naphthalene, tribenzopyrene, tetrabenzo(bgkP)chrysene, dinaphtho-pyranthrene-9,18-dione, 2-methyl-16-heptadecenoic acid, or dinaphthyl ketone. A refined model of the hexane soot emerged from this work and is shown in Figure 2.5.

The reduction of black carbon samples to finely divided forms can alter absorption spectra. Smith et al. (1975) point out that surface oxidation of the carbon materials can accompany the grinding in the presence of oxygen gas. They were able to develop absorption bands at

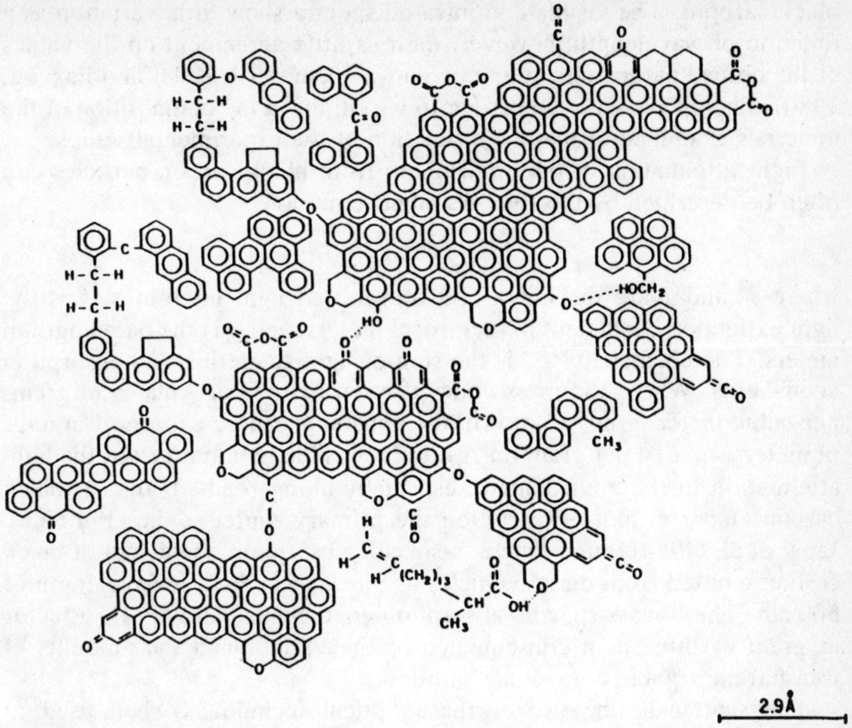

Figure 2.5. Representation of a hexane soot, with occluded substances, as formed in a flame (Akhter et al., 1984b). Reproduced with permission of the authors and the Optical Society of America.

1720, 1580, and 1240 cm^{-1} whose intensities depended upon the grinding times. These three frequencies could be assigned to known carbon–oxygen functional groups on the bases of previous studies: the band at 1720 cm^{-1} a ketonelike carbon; the band at 1580 cm^{-1} to a chelated carbonyl group; and the band at 1240 cm^{-1} to a simple carbon oxygen or phenoxy linkage. To confirm that these three frequencies are associated with carbon–oxygen functionalities, grinding of the carbons were carried out in nitrogen atmospheres. In such cases the absorptions did not develop.

OPTICAL PROPERTIES

Black carbons influence the passage of radiation by the atmosphere through their absorptive properties reflecting both absorptive and scattering effects. There is an extensive literature on the optical properties of black carbons. The visible and infrared spectra show little variation as a function of wavelength, however, there is little agreement on the values of the optical parameters at any given wavelength (Roessler and Faxvog, 1980). The differences can be due to variations in the composition of the materials examined or may be a function of the experimental setups.

Light attenuation in the atmosphere from black carbon particles can often be described by the Beer–Lambert equation

$$I/I_0 = \exp(-bx)$$

where I_0 and I are the initial and transmitted light intensities, b is the light extinction coefficient in reciprocal meters, and x is the pathlength in meters. The coefficient "b" is the sum of light scattering and absorption coefficients. Where the mass concentration of the particulates in grams per cubic meter (g/m^3) is known, the constants can be expressed in units of meters squared per gram (m^2/g). In the visible region much of the light attenuation in the environment, especially along roads, is due to black carbon (Japar et al., 1984). Often the primary source is diesel engines. Japar et al. (1984) found the mass-specific absorption coefficient of black carbon emitted from diesel vehicles as $9.8 \pm 1.5\ m^2/g$ at a wavelength of 500 nm. These mass-specific absorption coefficients can vary by a factor as great as three as a consequence of the variation in the amounts of transparent organic compounds emitted.

The scatter in the use of these optical techniques when used to determine particle mass concentrations results from variations in shape and particle chemistry (Roessler, 1984). Roessler used photoacoustic measurements to determine "b" for diesel exhaust particulates and showed that various phases of the fuel cycle produced various results. For example, at normal operating conditions, including idling, "b" was about $5\ m^2/g$, however it increased rapidly at higher load conditions and with low values of the air/fuel ratio.

The wavelength dependence of the extinction coefficient of three types of black carbon is shown in Figure 2.6 from the measurements of Blum and Fissan (1984). Such values of the extinction coefficients, based upon the integrating plate method (see Appendix), reflect absorption only and can be used only where scattering is unimportant. Blum and Fissan were able to estimate extinction coefficients of fire-produced particulates using Mie theory assuming spherical shapes.

The ability of black carbon particles to absorb radiation, compared to that of other particles, can be measured by the single scattering albedo, ω_0, which is the fraction of light scattered by a single particle. For most particles, the value is around 0.9–1.0 but for black carbons, since they

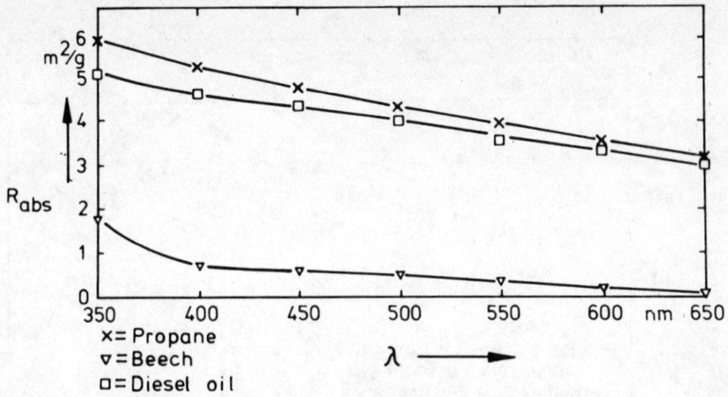

Figure 2.6. Absorption cross section per unit of particle mass, R_{abs}, as a function of wavelength for three test fires (Blum and Fissan, 1984). Reproduced with permission of the authors and Elsevier Science Publishers.

absorb light radiation effectively, have values close to 0.5. Chylek et al. (1984) indicate urban sulfate aerosols, which have single scattering albedos of between 0.6 and 0.7, require black carbon concentrations of around 5–17% by volume. On the other hand, the single scattering albedos of rural aerosols range between 0.8 and 0.9 and are consistent with concentrations of 0.7 to 5% black carbon. These theoretical results are in agreement with field observations.

The single scattering albedo has been modeled for two types of aerosols containing black carbons: external mixtures where the carbon and other aerosol components are mixed and internal mixtures where the black carbon is deposited on the outside shell of the aerosol or forms the nucleus of the aerosol (Bergstrom et al., 1982). The values of the single scattering albedos for sulfate and soil containing aerosols with differing amounts of black carbon are given in Figures 2.7a and 2.7b.

Roessler and Faxvog (1980) sought the attenuation coefficients, resulting from both scattering and absorption, for wavelengths of 0.5145 μm and 10.6 μm with soots generated from the combustion of acetylene. The results cannot be explained by the presence of small spheres using models based upon Mie theory. The agglomeration of the particles into clusters lowers the refractive index and the density of the soots. Further, the agglomerates are no longer spherical and the models are not exactly applicable. Although the visible spectral absorption is little affected by clustering, the infrared absorption is increased over that expected from small individual particles.

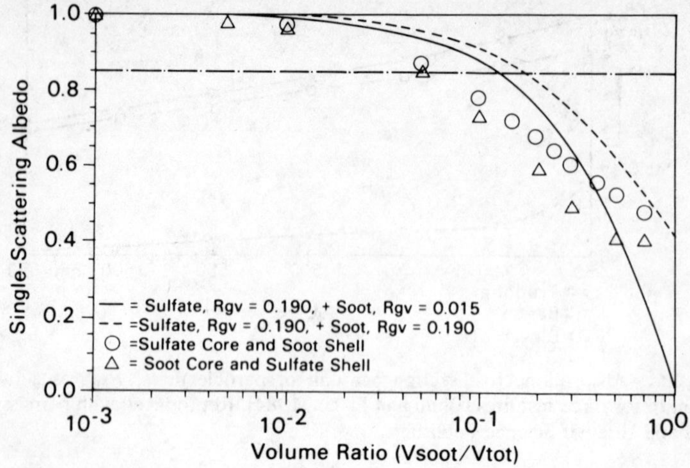

(a)

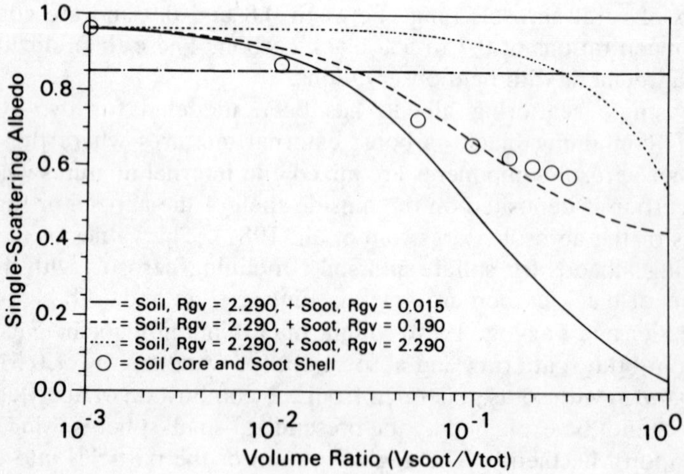

(b)

Figure 2.7. (*a*) Single scattering albedo as a function of the fractional volume of black carbon in a mixture of black carbon and sulfate. Rgv is the geometric mean radius of the particle. (*b*) Single scattering albedo as a function of the fractional volume of black carbon in a mixture of black carbon and soil material. (From Bergstrom et al., 1982). Reproduced with permission of the authors and Plenum Press.

ADSORPTIVE PROPERTIES

The property of black carbons to remove impurities from solutions and from gases has been known for many centuries. The types of charcoals used by different industries to clean up their products are identified by empirical approaches more than by objective assessments. Today, charcoals are used extensively in water purification processes. A review of the properties of active carbons is given by Mattson and Mark (1977) from which this summary is derived. The adsorption characteristics of black carbons have environmental consequences in the uptake of materials from their surroundings.

The porous nature of black carbons is in part responsible for their adsorption characteristics. The size of the pores is in the tenths of nanometers. The greater the porosity the greater the ability to transfer mass.

However, the surface chemical properties are of perhaps greater importance in determining the adsorptive capabilities of a given black carbon. The principle factors involved, according to Mattson and Mark (1977) are (1) the nature of the starting material; (2) the chemical characteristics of the environment in which the carbon was formed; and (3) the time and temperature of the reactions yielding the carbon. The activation process can start directly with biomass burning or with previously charred materials. Coconut shells produce a black carbon with a fine pore size distribution and with a high density. The types and amounts of minerals in the starting materials affect the absorptivity of the final product.

Industrially the activation process usually involves the heating of a partially oxidized starting material with a high carbon content to remove or decompose noncarbon impurities. If the process is carried out in the absence of oxygen, a subsequent oxidation process must follow to produce an active product. The oxidizing gas phase environment may possess O_2, CO_2, or H_2O as the active agents. Temperatures above 200°C are required. Solution oxidation usually involves conventional oxidizing agents, such as nitric/sulfuric acid solutions. The activation process always involves an oxidation of the surface carbons.

The nature of the active carbon sites has been extensively investigated. Their characteristics reflect those of oxygenated aromatic compounds. For example, there are such groups as carboxylic, enolic forms of 3-diketone, and phenolic as well as C–O and C–O–C bonds. The structural determinations, according to Mattson and Mark (1977), have not been very informative using conventional techniques.

Activated or active carbon has proven especially useful in the purification of waters from organic contaminants and in the isolation of such materials for analysis. These black carbons are prepared from chars through treatment with steam or carbon dioxide at temperatures between 800 and 1000°C to produce a variety of functional groups on the surfaces. BET surface areas as high as 2500 m^2/g have been measured. Pores as small as 4–10 Å in diameter have been observed (Andelman and Caruso, 1971). Both the nature of the starting materials and the characteristics of the charring and activation processes determine the absorption characteristics of the final product. Surface active groups include carboxyl, phenolic hydroxyl, and carbonyl, as well as some iron carbon centers developed from iron impurities in the starting materials. Such iron sites can be involved in redox reactions. Andelman and Caruso (1971) point out that such a property can be a nuisance in water control problems. Phenols form quinones on some surfaces. Some organic compounds can be destroyed and thus are not picked up in the analytical steps.

Absorption of organic solutes usually takes place according to Langmuir type isotherms. There will be competition set up between various solutes for any given system with pH, the presence of inorganic salts, and redox conditions affecting the actual absorptions. In systems with moving liquids the extent of solute absorption will be determined by the flow rate and diffusional processes.

For both chemical analysis and water purification, desorption processes achieve importance. The efficiency of desorption is a function of the extracting solvent and of the sorbed substances. Recoveries are often incomplete due to the very strong attraction of the solute to the carbon surface.

MORPHOLOGIES

There are three main sources of black carbons from combustion processes: coal, oil, and biomass burning. Natural gas combustion is less significant than that of other fossil fuels in the production of soots. In addition, black carbons are produced in gasoline and diesel engines of vehicles. The burning produces particles with characteristic morphologies (Griffin and Goldberg, 1979 and 1981). The shapes of the larger particles (>20 μm) can be categorized into three groups: (1) porous, spheroidal; (2) elongate, prismatic; and (3) irregular fragments. In addition, the surface textures can be characterized into three classes: (1) smooth, homogeneous; (2) rough, irregular with pits or cells; and (3)

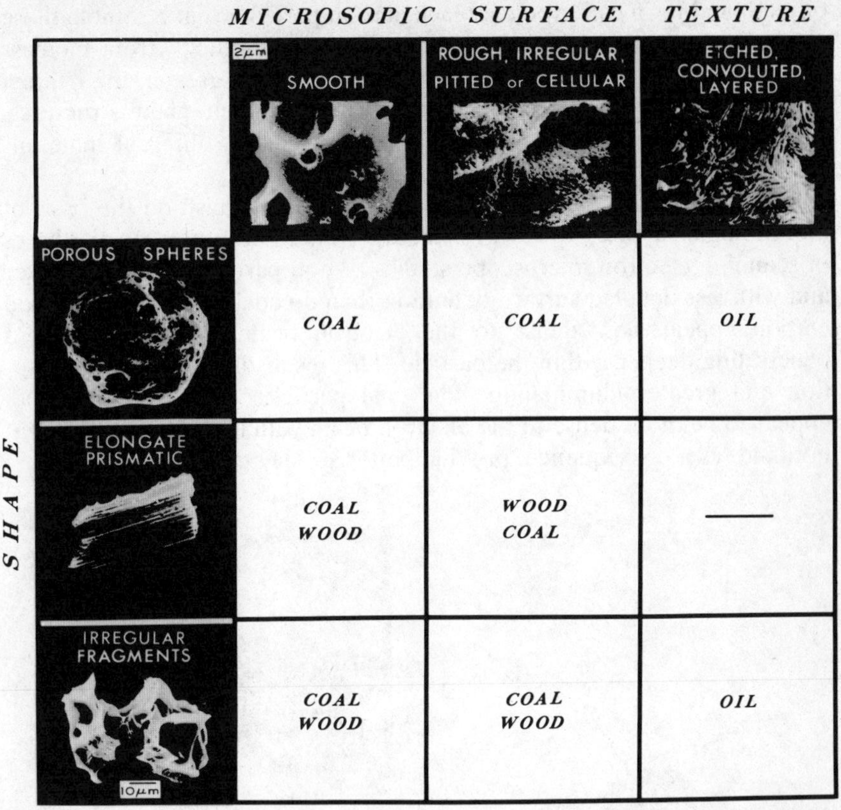

MICROSOPIC SURFACE TEXTURE
2μm
SMOOTH
ROUGH, IRREGULAR, PITTED or CELLULAR
ETCHED, CONVOLUTED, LAYERED
SHAPE
POROUS SPHERES
COAL
COAL
OIL
ELONGATE PRISMATIC
COAL WOOD
WOOD COAL
—
IRREGULAR FRAGMENTS
COAL WOOD
COAL WOOD
OIL
10μm

Figure 2.8. Classification of black carbon particles according to their shapes and surface texture. The shapes (vertical axis) are viewed under 1000× magnification and the surface textures (horizontal axis) are observed under 5000× magnification (Griffin and Goldberg, 1981). Reproduced with permission of the authors and Pergamon Press, Ltd.

etched convoluted layer structures. The coupling of these distinguishing features can be especially useful in determining the origin of black carbon particles found in environmental samples (Figure 2.8).

In general, the soots derived from oil burning are spherical (cenospheres) with delicate, convoluted, layered structures. Coal burning produces a variety of types, ranging from spheroidal particles with rough irregular pitted surfaces to elongate prismatic ones.

The charcoals from biomass burning (grass, brush, and forest fires) usually show cellular structures. The common morphology will be elongate prismatic with well preserved wood cells; nevertheless, irregular fragments are also produced. The burning of plants with large quantities of pitch and resin can produce black carbon particles that resemble those produced from coal burning. The black carbon particles from biomass burning commonly have ratios of length to width greater than three. Coal combustion particles very often have remnant plant structures. These characteristics of black carbons produced by different materials and burning processes are summarized in Figure 2.8.

Wood and coal carbons can be difficult to distinguish on the basis of morphologies alone. However, some differences are noted on the bases of scanning electron microscope studies. Wood particles appear brighter and with less detailed surface definition than do coal particles. The wood carbon appears less dense to the electron beam with the electrons penetrating deeper within the particle. This results in less surface resolution and greater illumination. The coal particles, on the other hand, appear to be more dense to the electron beam with less electron penetration and, as a consequence, provide better surface detail.

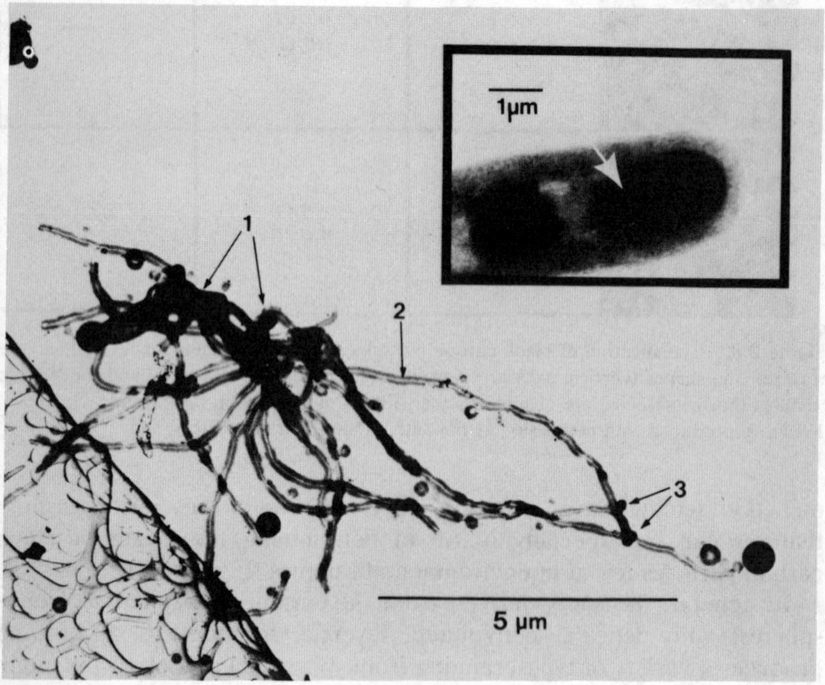

Figure 2.9 Transmission electron micrograph of filamentous carbon cluster. Arrow 1 indicates CuZn allow base of cluster. Arrow 2 points to hollow core structure of individual filaments. Arrow 3 indicates a CuZn alloy particle embedded at tip of an individual filament. Inset: Individual carbon filament with embedded CuZn particle (Buseck and Bradley, 1982). Reproduced with permission of the authors and the American Geophysical Union.

Other properties provide additional guidance to assigning origins to sedimentary charcoals. Samples with associated unburned coal particles or wood fragments would indicate combustion products from coal and wood burning, respectively.

Filamentous carbons, produced by the decomposition of carbon-containing gases such as carbon monoxide and hydrocarbons upon metal surfaces are introduced to the environment by the petroleum, petrochemical, nuclear, and mining activities (Busek and Bradley, 1982). Their formation is initiated by the decomposition of the gas upon a surface of a metal particle that adsorbs the carbon. The carbon diffuses to the other side of the particle where the filament forms (Baker et al., 1972).

In the refining of copper, the vessels with molten copper provide a surface for the growth of carbon filaments (Busek and Bradley, 1982). A strong thermal gradient exists above the vessel, which is apparently ideal for the growth of the filaments following the blowing of a hydrogen/hydrocarbon gas mixture through the metal melt. Large amounts of filamentous carbon are then discharged directly into the atmosphere.

In principle, these filamentous carbons possess a morphology to identify their source (Figure 2.9). However, as yet they have not been found in the environment.

SIZE DISTRIBUTIONS

Reliable measurements of size distributions of black carbons, either in the environment or issuing from their sources, are now being made as a consequence of improvements in analytical techniques. The developments of electrical aerosol analyzers, cascade impactors, and of scanning electron microscopic techniques have provided a basis for detailed studies of size distributions.

The small particle mode (those $<1\ \mu m$) most probably contain the black carbons formed through a condensation process involving gas phase moieties (see Chapter 3). The larger particle modes appear to be a combination of two or more discrete assemblages. Their sources can be the cenospheres and chars from fossil fuel and wood burning, and particles that have grown by the accumulation of other particles, be they black carbon, organic pollutants, or dissolved materials that mineralize (see below). Average particle size of black carbon aerosols in a given air mass increases with time (Heintzenberg and Covert, 1984; Berner et al., 1984). The importance of the submicron-sized particles has been emphasized in recent works. This class can penetrate deeply in the lungs, be involved in long-range transport, affect atmospheric visibility, and act as carriers for a variety of metals.

Multimodal distributions of black carbons issuing from flames, diesel engines, and freeway traffic have been found by Whitby (1979). The "nuclei mode" has a geometric mean diameter between 0.005 and 0.04 μm and probably results from the condensation of gaseous carbon moieties. The "accumulation mode" encompasses particles in the size range 0.15 to 0.5 μm and apparently results from the coagulation and condensation of the "nuclei mode" particles. Finally, in the case of vehicular emissions there is a "coarse mode" at several microns that is attributed to the precipitation of fine particles on the walls of exhaust systems and a subsequent entrainment in the issuing gases.

The coal-fired utility boilers produced soots with peaks at particle diameters approximately 0.1 μm (McElroy et al., 1982). One of the power plants investigated had the largest electrostatic precipitator known to be operating on a coal-fired installation. The precipitator was designed to remove fly ash at a 99.7% efficiency from the 520 MW boiler. The size distributions of the particles issuing from the boiler indicated the submicron mode and a larger particle mode with sizes >0.5 μm (Figure 2.10). The mass of the submicron mode is about one and a half percent

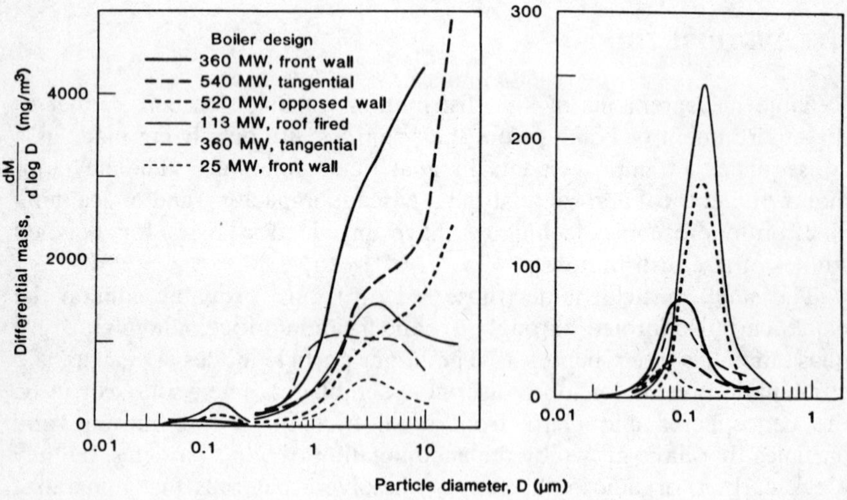

Figure 2.10. The size distribution of soots collected at the outlets of six utility coal-fired boilers (left). Expanded view of submicron size modes (right) (McElroy et al., 1982). Reproduced with permission of the authors and the AAAS.

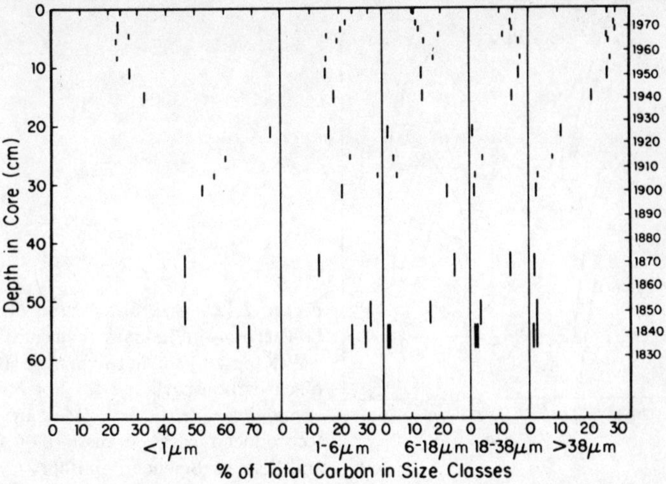

Figure 2.11. The size distributions of black carbon in Lake Michigan sediments as a function of time of deposition (Griffin and Goldberg, 1983). Reproduced with permission of the authors and the American Chemical Society.

of the total ash generated. The control devices alter this pattern substantially. The smaller particles are much more difficult to collect than the larger ones. In the plant stack emissions about 20% of the particulate mass was in the submicron class.

The size distribution of black carbons in Lake Michigan sediments varied as a function of deposition time (Figure 2.11). The material accommodated in strata before 1900 had between 50 and 75% of the total carbon in the submicron size class (Griffin and Goldberg, 1983). The small particles are attributed to an origin in forest fires in the midwest of the United States with a subsequent long range transport to the deposit site. In sediments deposited after 1900, the greater than 38-μm size class became dominant. The sources of these particles are the energy producing plants sited along the shores of lower Lake Michigan. The combustion of fossil fuel, combined with near-source fallout, accounts for the importance of the larger particles in recent sediments.

Long-range transport of the small size class is also seen in the black carbons of the Arctic haze. Sixty percent of the soot was in the less than 0.1-μm radius size class (Heintzenberg, 1982). The larger particles may have been preferentially removed from the air during its travels from sources in the industrialized areas of the lower latitudes.

The submicron particles have a relatively long persistence in the atmosphere and their size distributions may be relatively uniform from

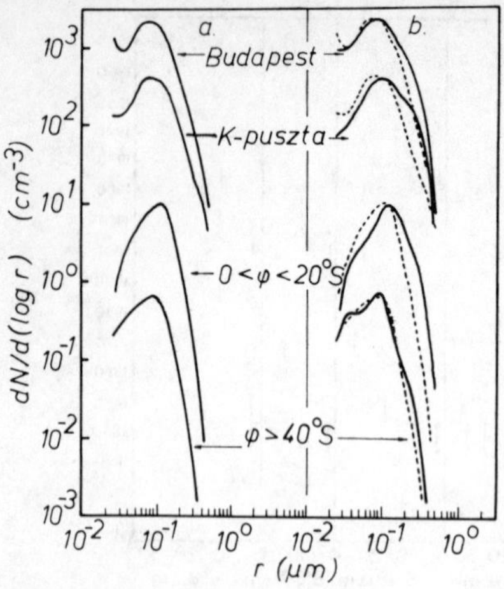

Figure 2.12. Size distribution of black carbon aerosols in Budapest, in a rural Hungarian site (K-puszta) and in two marine sites. (*a*) For black carbon particles. (*b*) For black carbon containing particles (Meszaros, 1984). Reproduced with permission of the author and Elsevier Science Publishers.

place to place. The size distribution in the range 0.2 to 0.5 μm for black carbons in aerosols taken from suburban Budapest, a rural station in Hungary, and in the South Atlantic were remarkably similar although the concentrations varied over nearly four orders of magnitude (Meszaros, 1984) (Figure 2.12). The data were obtained by electron microscopy.

Particle growth in the atmosphere through the accumulation of sorbed materials has been postulated for organic pollutants (van Vaeck et al., 1979). The smaller size fractions (<0.5 μm) were depleted in rural as opposed to urban aerosols although the total amounts of suspended solids were higher in the latter. The aging with growth of the aerosols formed in urban areas was proposed to account for this observation. There is some laboratory evidence on the adsorption of anthracene by ultrafine carbon aerosols to strengthen this hypothesis (Brown and Gentry, 1984). Acetylene soot particles exposed to anthracene concentrations of 1.3×10^{-8} g/cm^3 or greater at 100°C could raise the initial diameter of 14 nm to values as high as 19 nm.

The growth of inorganic crystals on black carbons from fuel oil combustion has been observed in the laboratory (Del Monte et al., 1984a). The particles after collection were exposed only to constant humidities and a remarkable variety of minerals were observed to grow on the cenospheres (Table 2.4). The elements in the crystal phases were supplied by the carbonaceous particles themselves. All of the particles are hydrated. The vanadium and nickel probably had origins in the combustions of the porphyrins associated with the oil.

Table 2.4. The Sequence of Crystal Growth on Carbonaceous Particles from Oil Combustion (Del Monte et al., 1984)[a]

Growth Order	Elemental Symmetry	Chemical Composition[b]	Composition
1	Prismatic hexagonal	Na V	$NaVO_3 \cdot 3.5H_2O$
2	Prismatic hexagonal	Na V	$NaVO_3 \cdot 1.9H_2O$
3	Prismatic hexagonal	Na <u>S</u> V <u>Fe</u>	Undetermined
4	Tetragonal bipyramidal	Mg K <u>S</u> V <u>Ni</u>	Undetermined
5	Monoclinic twinned	Na Mg S <u>V</u> Fe <u>Ni</u>	Undetermined
6	Monoclinic twinned	Na <u>S</u> <u>V</u>	Undetermined
7	Monoclinic	S <u>V</u> <u>Fe</u>	Undetermined
8	Amorphous	Si	$SiO_2 \cdot nH_2O$
9	Monoclinic	S Ca	$CaSO_4 \cdot 2H_2O$

[a]Reproduced with permission of the authors and Elsevier Science Publishers.
[b]The principal elements are underlined.

3

BLACK CARBON FORMATION

INTRODUCTION

Different forms of black carbon are created during the combustion of organic materials. Four distinct types have been identified by Medalia and Rivin (1982). The smallest particles are formed through deposition from the gas phase and have been designated by Medalia and Rivin (1981) as "aciniform carbon." They are spherical or near-spherical with mean diameters around 20 to 30 nm and they adhere to each other to form straight or branched chains (Wagner, 1978; Figure 3.1). There are about a million carbon atoms in each of the spheres. Aciniform is defined as "clustered like grapes."

A second form encompasses the cenospheres—hard, shiny, porous, or hollow carbon spheres—which are typically of 10–100 μm diameter. They form when liquid drops undergo carbonization without major changes in shape. They are produced from coal and oil burning.

A third form includes the charcoals or chars—small fragments of carbonized woods or coals—ranging in size from microns to fractions of meters. They often maintain the structures of plant, tree, or animal parts from which they originated.

Finally, Medalia and Rivin recognize the "carbonaceous microgels," which are formed from microscopic entities in which spheroidal carbon particles with colloidal dimensions are imbedded in organic materials.

The C-14 activity of black carbons in principle can reveal whether or not the material was produced from modern or old sources. Coals and oils contain no primary C-14 activity, whereas contemporary carbon-containing materials such as trees or plants would have that of atmospheric carbon dioxide (Currie, 1982).

Cooper et al. (1981) measured the C-14 activity of milligram amounts

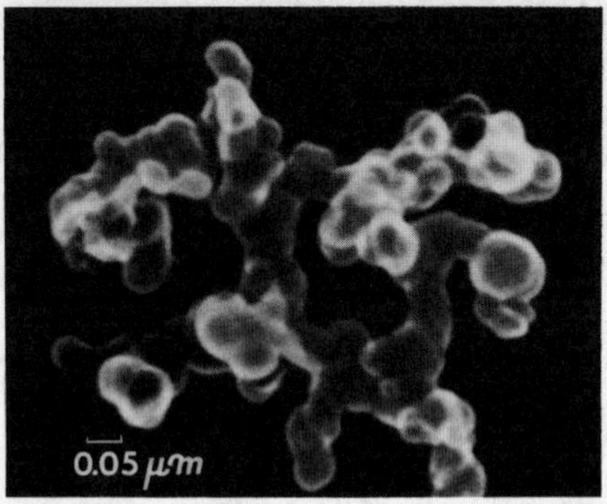

Figure 3.1. A small black carbon particle as seen under the transmission electron microscope.

of atmospheric particulates collected from Oregon airsheds. They extracted the carbon dioxide from the particles whose large size fraction included pollen, spores, woods fibers, and so on, as well as the under 2-μm size fraction, which probably contained the bulk of the black carbon. In this part of the U.S. west coast, wood burning is especially prevalent during the winter periods and the total suspended particulates reflected this with high C-14 activities. Slash and field burnings also took place.

Woods that have formed over the life of the tree give an apparent age reflecting contributions from organic phases deposited during this period. There are further perturbations upon the determination of a real age from biological material as a consequence of increased levels of C-14 in the atmosphere from nuclear weapons testing during recent times and increased levels of inactive carbon dioxide from fossil fuel burning.

HIGH TEMPERATURE FORMATION

The formation of black carbon particles should be initiated when m in the following equation exceeds the value of $2y$

$$C_mH_n + yO_2 = 2yCO + (n/2)H_2 + (m - 2y)C$$

Wagner (1978) points out that experimentally, the particles do not form when C/O = 1, but only when the ratio is about 0.5. Carbon dioxide and water are found in the product gases; thermodynamically, they should not form along with the soot. Reaction kinetics are responsible for charcoal formation at low values of the ratio. A schematic mechanism is proposed:

$$RH + OH \xrightarrow{1} R_i + H_2O$$

$$R_i + OH \xrightarrow{2} RCO + R_h^\circ$$

$$CO + OH \underset{-3}{\overset{3}{\rightleftharpoons}} H + CO_2$$

$$H_2 + OH \underset{-4}{\overset{4}{\rightleftharpoons}} H + H_2O$$

$$R_h^\circ + R_nH \xrightarrow{5} R_{n+h}H$$

The reactions with hydrogen atoms (−3 and −4) have high activation energies. Thus the production of water and carbon dioxide depends upon the relative rates of the reactions that consume the fuel, CO and H_2.

The reactant from the high temperature reactions polymerizing to black carbon is generally accepted to be acetylene, which appears to be the last remaining hydrocarbon before solids begin to form. Elemental carbon does not appear as C_2 in band spectra observed at the lower, solid forming, temperatures. In a similar way, benzene and other aromatics are dismissed as potential reactants as their absorption spectra disappear well before carbon luminosity is evident. On the other hand, acetylene is formed at 1300 K and persists till 2300 K, the temperature range at which solid carbons are formed.

There appears to be a polymerization of the acetylene molecules and a dehydrogenation, processes that may go on simultaneously and in a coupled way. Abrahamson (1977) suggests that there are probably larger intermediates formed from the acetylene before the solid carbons are formed. In the formation of carbon from acetylene, the particles continue to grow, even after all of the acetylene is used up. Mass spectrometric studies of molecules produced in the acetylene combustion indicate that polyacetylenes and then larger molecules in the mass range 300–500 were formed before solid carbons appeared. The latter were a much lower concentration and Abrahamson (1977) indicates that there is a direct route through polyacetylenes to solid carbon. The particles collected earlier had C/H ratios as high as one.

Abrahamson (1977) postulates the formation of platelets, initially with an H/C ratio of one, which he designates as "saturated platelets." These platelets can maintain radical sites and acetylene can add on to them by radical reactions. Graphite layers build up from the inside and hydrogen is lost as H_2. Dehydrogenation and acetylene uptake thus both occur simultaneously. If the former process is more rapid than the latter, then polycyclic aromatics, which are observed in the flames and in the resultant black carbon (see below), can be produced. Observed by electron diffraction in the first solids formed are the 0.154-nm singly bonded C–C bonds and 0.141-nm graphitelike spacings. Further, these first particles have 100 times more electron resonance spins than do the older particles, indicating the presence of radical sites.

The formation of the carbon spheres is perhaps mediated by an electrical mechanism, according to Abrahamson (1977). He proposes that a weak plasma exists in the pyrolysing hydrocarbons and contains small positive ions, large negative platelet ions and neutral particles. Some of the neutrals will be platelets and some of the platelets and smaller molecules will be radicals. As a consequence, electrical-radical nucleation will occur:

1. Some radicals will attach ions, usually the smaller, more rapidly moving positive ions, producing positive radical ions.
2. The positive radical ions will attract negative platelets.
3. The neutral positive radical ion–platelet combination will attract positive radical ions by collisional processes.
4. The positive radical ion, much larger in size, will now attract a negative platelet forming a neutral species.

In such a way nucleation ends and the buildup of a carbon sphere begins. The validity of this hypothesis is strengthened by observations of the physical form of the carbon balls. The basal layers are aligned parallel to the surface, perhaps suggesting that there is a wrapping around of platelets deposited from the gas phase. The interlayer spacings are well in excess of those of graphite. On the other hand, thermally produced carbon blacks are uniform spheres with a center of growth. Furnace and channel blacks show many growth centers, embedded together by more onionlike layers on the outside. The different physical forms of soot, Abrahamson (1977) argues, may relate to the electrically aided process and the factors that control the interaction of the species in the plasma.

The details of the surface growth have been examined by studies of premixed flames by Harris and Weiner (1983a, 1983b). The increased surface growth of black carbons following nucleation is first order in the acetylene concentration, the growth species. This can be explained by an increase in surface area available for growth rather than by an increase in the concentration of the acetylene.

Following the creation of the black carbon particles in the nucleation phase (1–2 nm), the heterogenous process of growth takes place (Harris and Weiner, op. cit.). The cessation of growth is a consequence of the decrease in the reactivity of the soot surface, which may be determined by its C/H ratio, its radical character, the temperature, or its age, but *not* its size.

LOW TEMPERATURE FORMATION

There is evidence that environmental black carbons may form at temperatures much below those at which a flame is evident. Experiments usually depend upon an identification based on X-ray diffraction spectra of reaction products. Kerogens, from an estuarine complex in coastal Louisiana, were heated at 300, 600, and 900°C successively for 50 hr at each temperature. There was a progressive development of graphite diffraction lines with increasing temperature (Harrison, 1976). The original material contained woody fragments as well as amorphous materials and in it there was no evidence of the sensitive d_{002} graphite reflection that became more and more developed from heating. The kerogen does order to a graphite structure at temperatures as low as 300°C. Higher temperatures develop higher-order crystallinities. The original composition of the starting materials was 62.1% C; 5.2% H; 29.2% O, and 3.5% N.

ASSOCIATION OF CHARCOALS WITH OTHER SUBSTANCES

There are a large number of inorganic and organic compounds accompanying the formation of black carbons in fossil fuel and wood burning, many of which may be associated directly with the carbon particles (Daisey, 1980). A major concern involves the observation that many of the compounds are carcinogenic. This sense developed over 200 years ago when Sir Percival Pott indicated a relationship between the high incidence of scrotal cancers among chimney sweeps and their exposure to soots. Recently, the carcinogen dioxin has been found associated with soots collected from mufflers at power houses and from fireplaces (Long and Hansen, 1983).

Polycyclic aromatic hydrocarbons (PAH) accompany soots and charcoals. In benzene soots there is evidence for polymeric PAH structures (Klempier and Binder, 1983).

The incomplete combustion of carbonaceous materials produces not only black carbons but also members of the arene class of hydrocarbons, condensed polynuclear aromatic hydrocarbons where all but one of the benzene rings is condensed. These arenes, along with some other organic compounds, are characterized by an extractability into benzene.

The greater portion of these particles occur in the less than 3.5-μm size class which is respirable. For example, over 90% of the total amount of benzo(a)pyrene is found in particles <0.2 μm in diameter. Many of the polynuclear aromatic compounds formed during combustion condense upon the combustion particulates after they cool (Natusch, 1976). Investigations have involved both fly ash, atmospheric particulates, industrially produced charcoals, and emissions from furnaces and engines.

Power Plants

The emitted solids from coal- and oil-fired power plants, the largest anthropogenic sources of particulates to the atmosphere, have been characterized on the basis of color (Cheng et al., 1976) and upon the basis of chemical and physical properties (Hulett et al., 1980). Soot may act as a tracer for these particles in their dispersion about the atmosphere and in their subsequent settling to the earth's surface and accommodation in the sedimentary record. These two investigations provide guidance as to what minerals and what elements might accompany the soot.

Cheng et al. (1976) categorized the particles on the basis of color as observed under the light microscope. Particles from an oil-fired burner were placed into three groups: black, white, and colored. The black particles, often with white and colored particles in aggregation, are generally carbon-containing, porous spheres. The white particles are irregular crystals or clusters of spherical particles containing silicon. The colored particles contained metals and sulfur in the forms of salts and oxides. Particles from a coal-fired burner ranged in color from milk white through yellow, grey, orange, brown to black. They were spherical with rather smooth surfaces, similar to those found by Griffin and Goldberg (1981) for power plant emissions. The black spheres were thought to be unburned hydrocarbons, but they most probably are black carbon.

Over 90% of the particles from both sources were spherical, but particles from the oil power plant were a bit more irregular. The associated elements with the particles from the power plants were as follows:

1. Particulates from oil-fired burner.
   a. Black cenospheres.

Major elements	Si, S, Ca, V, and Fe
Minor elements	Mg, Al, P, Ti, Cr, Mn, and Ni

b. White particles.

Major elements	Si
Minor elements	Al, S, K, and Ca

c. Colored particles.

Major elements	S and Fe
Minor elements	Ni, V, and Mg

2. Particulates from coal-fired boiler.
   a. Glass spheres.

Major elements	Si, S, K, Ca, Ti, and Fe
Minor elements	Al, P, Cl, Mn, and Cu

Fly ash from four coal-fired power plants have been separated into three categories by a combination of size and magnetic fractionations with chemical treatments (Hulett et al., 1980): glass, mullite-quartz, and magnetic spinel. The silicate particles were sintered to the magnetic materials, making complete separations by physical means not possible. The magnetic materials were dissolved in HCl leaving the alumino-silicate minerals as a residue. The glass materials were separated from the quartz and mullite by dissolution in 1% HF. The spherical nature of some of the particles, containing laths of mullite, is indicated by the skeletons produced by the HF treatment wherein the interstitial glasses were removed.

The mullites had compositions approximating that of the natural mineral, $3(Al_2O_3) \cdot 2(SiO_2)$. The quartz phases had usually high aluminum concentrations, ranging between 2 and 10% by weight. The high concentrations of aluminum in the magnetic materials suggested that the spinels are ferrite rather than magnetite. The material has the composition of $Fe_{2.3}Al_{0.7}O_4$. Aluminum oxide, Al_2O_3, was not present but about 10% of the magnetic phases were α-Fe_2O_3.

Elemental distributions among the three groups are shown in Table 3.1 for one of the fly ashes studied. Cr and Ga were consistently more abundant in the mullite-quartz phases than in the glasses for the fly ashes studied. For these nonmagnetic components, the alkali, rare earth, and transition elements prefer the glass phases to the mullite-quartz. With the exception of Mg and Ca the monovalent and divalent metals are more

Table 3.1. Element Distributions (ppm) in the 100- to 200-μm Fraction of Bull Run Fly Ash (Hulett et al., 1980)[a]

Element	Glass	Mullite-Quartz	Magnetic Spinel (HCl Extracted)	Element	Glass	Mullite-Quartz	Magnetic Spinel (HCl Extracted)
Al	87,200	204,000	86,600	Ga	0	12	0
Fe	19,600	3400	605,000	As	42	2	0
Si	343,000	287,000		Se	54	7	38
Na	1710	256	2080	Mo	0	3	0
K	25,000	5270	3590	Cd	0.32	0.27	1.2
Rb	532	12	229	Hf	6	3	0
Cs	0	0	0	Ta	3	1	0
Mg	10,400	15,500	0	W	4	1	0
Ca		0	0	Hg	8	1	0
Sr	798	174	0	Pb	35	2	
Ba	319	69	0	La	61	0	14
Sc	31	7	15	Ce	119	6	141
Ti	10,900	4630	0	Sm	0	0	0
V	176	127	150	Eu	2	0	1
Cr	169	311	1380	Tb	5	0	0
Mn	87	9	1040	Dy	10	0	0
Co	25	2	251	Yb	4	0	0
Ni	174	78	2270	Th	5	0	5
Cu	144	17	496	U	13	1	4
Zn	209	17	382				

[a]Reproduced with permission of the authors and the American Association for the Advancement of Science.

readily accommodated in the glasses. Associated with the magnetic phases are Cr, Mn, Co, Ni, Zn, Cu, Cd, and Ce. The high concentrations of some alkali metals in the HCl extract of the magnetic phases are attributed by Hulett et al. (1980) to the presence of soluble silicates.

Some of the metals with valences of three or four are highly concentrated in the mullite-quartz, Ti, Cr, and so on, and probably substitute isomorphically for Al or Si. Similarly, the elements that are enriched in the spinels probably are in the form $Fe_{3-x}M_xO_4$, where M designates the metal.

The use of mullite as a tracer for anthropogenic coal burning has been previously proposed (Smith et al., 1973). The association of mullite and black carbon in sediments clearly is evidence for the entry of fly ash. Also, the presence of spinels, as opposed to magnetite, in the magnetic fractions of sediments is further confirmatory evidence. The entry of metals to sediments clearly can be correlated in principle with the entry

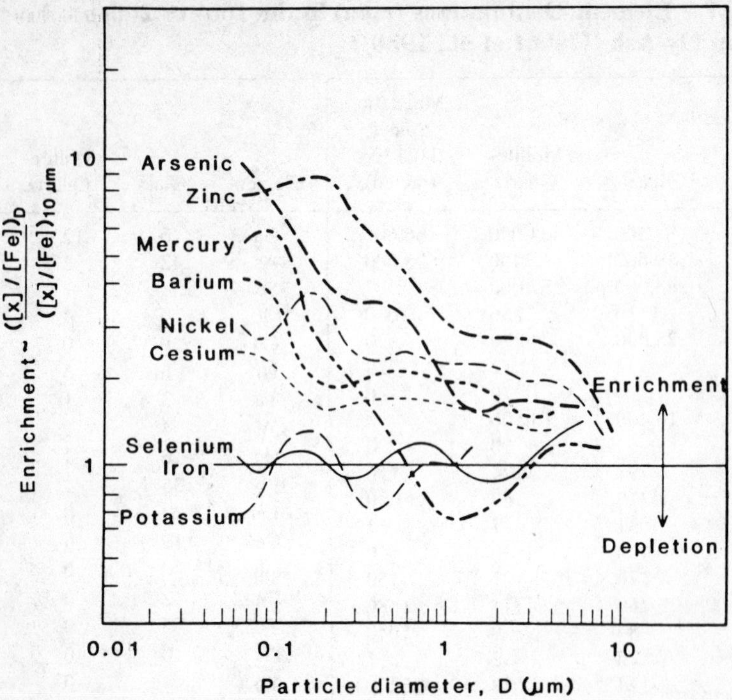

Figure 3.2. Enrichments of trace metals in emissions from coal-burning boiler outlets. Enrichment is defined as the concentration ratio of the element to iron at a specific particle diameter divided by the ratio at 10 μm (McElroy et al., 1982). Reproduced with permission of the authors and the American Association for the Advancement of Science.

of these and other fly-ash components (see Lake Michigan study in Chapter 8).

The metals issuing from boiler outlets of coal-burning power plants are continuously distributed over the entire particle size range and have concentration profiles that are similar to those of the total mass (McElroy et al., 1982). Essentially all of the elements display a peak in the submicron mode (Figure 3.2). This result suggests that these metals may have volatilized during the coal burning and associated with the smelter particles, perhaps including those of black carbon. The enrichment factor (EF) of the metals in various size classes is indicated by the concentrations of element x and of iron in size fractions of diameter D and of 10 μm:

$$EF = \frac{(x)/(Fe)_D}{(x)/(Fe)_{10\ \mu m}}$$

From Figure 3.2, the more readily volatilized arsenic, zinc, and mercury appear to be enriched in the smaller size fractions.

The long-range transport of metals clearly favors those associated with the smallest size classes. For example, Pb-210 has been observed to have 70–90% of its activity in the aerosols, whose diameters are 1.2 μm. Collection was made at open ocean sites (Sanak et al., 1981).

Fly-ash particles collected from a coal-burning plant were divided into two groups for surface studies (Hock and Lichtman, 1982): those that are electrically conductive and constitute about 10% of the material and those that are electrically insulating. The former consist mostly of black carbon while the latter contain a variety of elements. The particles in the size range 5–30 μm, over 90% of which are spherical, were studied by electron microscopy and Auger electron spectroscopy.

The noncarbon, nonconducting particles had a very thin (<1 mm) surface layer of carbon, which the investigators suggested was perhaps a hydrocarbon and which was readily removed by ion sputtering. In this layer were also S and Cl. The carbon concentrations fell markedly below the surface from 29% of the particles to 5.1% at a depth of 1 nm. There was a negative correlation between oxygen and carbon in the surface layer. The electrically conductive particles consisted of carbon at least to a depth of 30 nm. The carbon was most probably in the elemental form.

Particulate emissions from a coal-fired heating plant contained dimethyl sulfate and monomethyl sulfate in concentrations as high as 839 ppm (Lee et al., 1980). The latter compound is presumed to come from the former through hydrolysis. The dimethyl sulfate decomposes at its boiling point so that it must be formed downstream of the combustion chamber. It can then become associated with the fly-ash particles. As yet there have been no studies of the formation of the dimethyl sulfate with oil burning. Both the parent and daughter compound are known to have mutagenic and carcinogenic properties.

Aerosols

The particulate organic matter (POM) in aerosols has been categorized by Daisey (1980) into the following groups:

1. *Hydrocarbons*. These are the alkanes, alkenes, and some aromatics with the aliphatics constituting the greatest fraction. They range from C_{17}–C_{37}.
2. *Polycyclic aromatic hydrocarbons*. These are the most intensely

studied set, primarily as a consequence of the carcinogenicity of some of the members.

3. *Oxidized hydrocarbons*. These classes include acids, aldehydes, ketones, quinones, phenols, and esters, as well as the less stable epoxides and peroxides. They may be produced directly in combustion processes or through oxidation reactions in the atmosphere.

4. *Organo-nitrogen compounds*. The aza-arenes are the only types of this class that have been so far analyzed and they are one or two orders of magnitude less than the polycyclic aromatic hydrocarbons.

5. *Organo-sulfur compounds*. Heterocyclic sulfur compounds, such as benzothiazole have been reported in urban aerosols.

The covariance of polycyclic aromatic hydrocarbons (PAH) and black carbon concentrations in aerosols has been demonstrated in studies carried out in Sweden (Björseth et al., 1979) where samples were collected near Gothenburg during February, 1977. The maxima and minima were in concord for air masses that presumably had traveled from Great Britain:

February 1977	Black Carbon ($\mu g/m^3$)	PAH (ng/m^3)
3–4	41.3	92.4
5–7	14.5	14.3
9–10	6.3	6.7
4–16	12.7	7.6
9–21	11.2	23.4
4–26	8.8	8.2

Industrially Produced Black Carbons

The industrially produced black carbons are quenched to impart distinctive characteristics in particle size, surface activity, and chain structure, which are demanded for their commercial usage. On the other hand, combustion processes are carried out to maximize energy production with little or no quenching. The industrial black carbons are purer than their combustion counterparts. The latter are larger particles and incorporate larger quantities of benzene extractable materials and they have a higher volatile component than do industrially produced soots

Table 3.2. Compounds Identified in the Extract of Carbon Black (Peaden et al., 1980)

1. Pyrene
2. Benz[a]anthracene
3. Benzo[b]fluoranthene
4. Perylene
5. Benzo[k]fluoranthene
6. Benzo[a]pyrene
7. Benzo[ghi]perylene
8. Indeno[1,2,3-cd]pyrene
9. Benzo[b]perylene
10. Dibenzo[cd,jk]pyrene
11. Coronene
12. Tribenzo[bc,hi,mno]aceanthrylene
13. Dibenzo[a,c]pyrene
14. Dibenzo[cd,lm]perylene
15. Acenaphto[1,2-k]fluoranthene
16. Tribenzo[a,cd,jk]pyrene
17. Dibenzo[b,pqr]perylene
18. Naphtho[1,2,3,4-ghi]perylene
19. Naphtho[8,1,2-bcd]perylene
20. Dibenzo[a,ghi]perylene
21. Diindeno[1,2,3-cd:1′,2′,3′-jk]pyrene
22. Benzo[a]coronene
23. Benzo[cd]naphtho[5,4,3,2-fghi]perylene
24. Phenanthro[5,4,3,2-efghi]perylene
25. Benzo[1,2]pyreno[3,4,5,6-jklma]dibenzothiophene
26. Benzo[9,10]pyreno[3,4,5,6-jklma]dibenzothiophene
27. Dinaphtho[8,1,2-bcd:2,1,8-jkl]pyrene
28. Dinaphtho[2,1,3-cde:2′,1′,8′-jkl]pyrene
29. Coroneno[2,3,4-bcd]benzothiophene
30. Tribenzo[2,1,4,5,6,7]chryseno[10,11-bcd]thiophene
31. Naphtho[5,4,3-abc]coronene
32. Dibenzo[n,pqu]indeno[1,2,3-ghi]perylene
33. Phenanthro[7,6,5,4b,4a,4:12,12a,12b,1,2,3]-peryleno[6,7-bcd]thiophene
34. Dibenzo[a,cd]naphtho[8,1,2,3-fghi]perylene
35. Dibenzo[ij,rst]naphtho[6,5,4,3-defg]pentaphene
36. Benzo[ij]dinaphtho[6,5,4,3-defg:6,5,4,3,2-pqrst]-pentaphene
37. Benzo[a]naphtho[5,4,3-ghi]coronene
38. Ovalene
39. Benzo[a]naphtho[5,4,3-def]coronene
40. Dibenzo[de,kl]phenanthro[2,1,10,9,8,7-pqrstuv]-pentaphene
41. Benzo[ghi]dinaphtho[8,1,2-bcd:8,1,2-klm]perylene
42. Benzo[a]naphtho[2,1,8-efg]coronene

Table 3.2. (*Continued*)

43.	Benzo[a]naphtho[2,1,8-hij]coronene
44.	Phenanthro[1,10,9-abc]coronene
45.	Phenanthro[9,8,7-abc]coronene
46.	Dinaphtho[5,4,3-abc:2,1,8-klm]coronene
47.	Dinaphtho[5,4,3-abc:5,4,3-ghi]coronene
48.	Benzo[a]ovalene
49.	Benzo[d]ovalene
50.	Dibenzo[fg,kl]phenanthro[2,1,10,9,8,7-pqrstuv]-pentaphene
51.	Pyreno[3,4,5-abc]coronene
52.	Dibenzo[fg,ij]phenanthro[2,1,10,9,8,7-pqrstuv]-pentaphene
53.	Pyreno[4,3,2-abc]coronene

Reprinted with permission of the authors and the American Chemical Society.

(Thomas et al., 1968). For example, the highly impure furnace oil blacks have 0.05% benzene extractable substances (Smith, 1964).

Thomas et al. (1968) indicate the simultaneous formation of black carbon and arenes, especially the well studied and abundant benzo(a)pyrene, B(a)P. They suggest a hypothetical black carbon molecule that is composed of platelets made of extensions of coronenes. For example, 91 fused rings result in a $C_{216}H_{36}$ molecule that yields a substance with a chemical composition similar to that of soots. During soot formation, oxidation can occur on the surface resulting in a surface covering of 30% with an incorporation of 5 to 10% oxygen. This, of course, is the activated carbon. The arenes, such as B(a)P, are presumed to be adsorbed on the surface of the soot through hydrogen bonding. This type of attachment then provides for ready extraction into solvents such as benzene.

Thomas et al. (1968) indicate that the preferential formation of black carbon depends upon the presence of substances that reduce the concentration of hydrogen atoms, while the introduction of hydrogen to the combustion chamber should enhance the production of arenes. Experimentally, when hydrogen is added to a propane flame, there was an increased production of B(a)P and of benzene extractable organics. On the other hand, the addition of nitropropane, a radical sink for hydrogen atoms, to the propane gas flame results in a reduction of pentane soluble material in the soot as well as of B(a)P. When air flow is increased to flames, the production of both soot and extractables decreases and combustion becomes a more nearly stoichiometric process.

High molecular weight (>300 daltons and containing up to 11

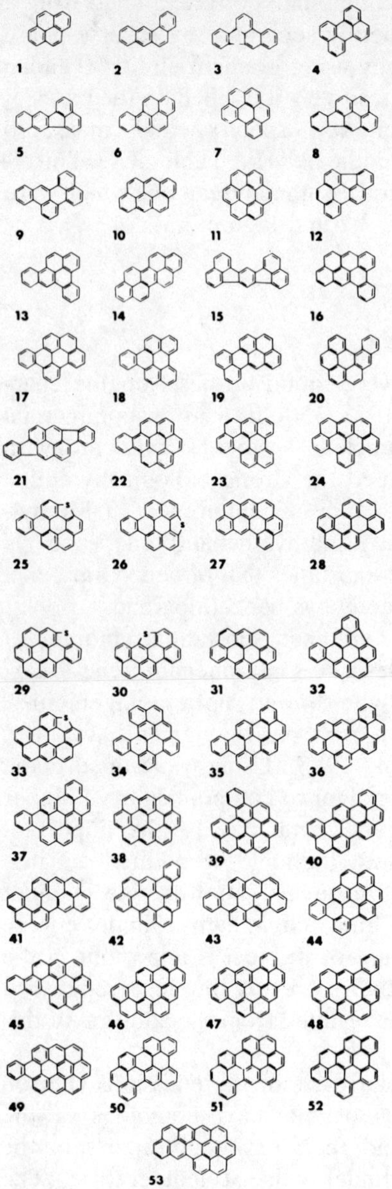

Figure 3.3. Structures of compounds in an extract of carbon black. The numbers are the same as those in Table 3.2 (Peaden et al., 1980). Reproduced with permission of the authors and the American Chemical Society.

condensed rings) polycyclic aromatic compounds were extracted from a commercial carbon black and subsequently separated by reverse-phase high-performance liquid chromatography using gradient elution (Peaden et al., 1980). Fifteen compounds were positively identified on the bases of molecular weight, fluorescence excitation spectra, and measured chromatographic retention data are indicated in Table 3.2. Thirty-eight others were given tentative structural assignments, among which were several large sulfur heterocycles (Table 3.2 and Figure 3.3).

Diesel Engines

A number of mutagenic compounds were found in diesel engine charcoals collected from an Oldsmobile (1978 350CID run on commercial fuel) by Yu and Hites (1981). Polynuclear aromatics extracted with hexane and toluene (Table 3.3) contained the strong biologically active compounds, the alkylphenanthrenes and alkylfluorenes. 1-Methylphenanthrene, 2-methylphenanthrene, 9-methylphenanthrene, and 9-methylfluorene have mutagenicities 0.3–0.5 times that of benzo(a)pyrene, while 1,9-dimethylfluorene is twice as active as benzo(a)pyrene.

The fraction isolated with toluene contained substantial amounts of aromatic ketones and aldehydes (Table 3.3). Only phenanthrene-9-carboxaldehyde and pyrene-1-carboxaldehyde showed slight mutagenicities.

The mutagenicities of soot emitted from diesel engines have been related to nitropyrenes (Henderson et al., 1983). Using triple-quadrupole mass spectrometry they found that the polynitro compounds have higher mutagenicities than the mononitropyrenes and may be the important toxic agent in soots emitted from diesels. It should be pointed out that only two diesel engines were used in the investigation (a one cylinder engine from the People's Republic of China and an eight cylinder engine from the United States). Also, the nature of the fuel is important in the characteristics of the emissions. The fuel used was low in pyrenes, but only minor differences in mutagenicities resulted from the addition of this chemical to the diesel oil.

Subsequent work developed the systematics of the emissions (Jenson and Hites, 1983). The concentrations of the alkyl homologs of the polynuclear aromatic hydrocarbons and their oxy counterparts in the particulates varied inversely with cylinder exhaust temperature. The degree of alkylation for the most abundant homolog of these compounds increased by one or two carbons with decreasing cylinder exhaust temperature.

The polynuclear aromatics associated with the black carbon emissions

Table 3.3. Compounds Found in the Hexane/Toluene Fraction of the CH_2Cl_2 Extract of Diesel Exhaust Particulates (Yu and Hites, 1981)

Methylfluorenes
Dibenzothiophene[a]
Phenanthrene
C_2-Fluorenes
3-Methylphenanthrene
2-Methylphenanthrene
9- and 4-Methylphenanthrene
1-Methylphenanthrene
C_3-Fluorenes
Phenylnaphthalene
C_2-Phenanthrenes
Fluoranthene
C_4-Fluorenes
Pyrene
Methylphenylnaphthalenes
C_3-Phenanthrenes
Methylfluoranthenes and methylpyrenes
C_2-Phenylnaphthalenes
C_4-Phenanthrenes
C_3-Phenylnaphthalenes
Benzo[ghi]fluoranthene[a]
Benzo[a]anthracene
Chrysene or triphenylene
Nitropyrene or nitrofluoranthene
Benzofluoranthenes
Benzopyrenes
Perylene
2-Naphthaldehyde
1-Naphthaldehyde
Methylnaphthaldehydes
6-Methyl-2-naphthaldehyde
Biphenylcarboxaldehydes
Fluorenone
C_2-Naphthaldehydes
Methylbiphenylcarboxaldehydes
Methylfluorenones
C_3-Naphthaldehydes
2-Methylfluorenone
C_2-Biphenylcarboxaldehydes
C_2-Fluorenones
9,10-Anthracenedione
4H-Cyclopenta[def]phenanthren-4-one[a]

Table 3.3. (*Continued*)

Methyl-9,10-anthracenedione
C_4-Naphthaldehyde and methyl-9,10-anthracenedione
Phenanthrenecarboxaldehydes
2-Phenanthrenecarboxaldehyde
C_3-Fluorenones and C_4-fluorenones
C_5-Fluorenones
Methylphenanthrenecarboxaldehydes
C_5-Fluorenones
C_6-Naphthaldehydes
Benzanthrone/11-benzo[a]fluorenone[a]
Benzo[b]naphtho[2,1-d]thiophen/other isomers[a]
7H-Benz[de]anthrone-7-one[a]
Pyrenecarboxaldehyde or fluoranthenecarboxaldehyde

[a]Tentative identification.

Reprinted with permission of the authors and the American Chemical Society.

from diesel engines primarily arise from their incomplete combustion in the fuel of which they were components (Henderson et al., 1984). The burning of pure hexadecane yielded soots with much lower mutagenic responses to *Salmonella* than that of reference diesel fuels in a People's Republic of China Swan diesel engine. Pyrene and phenanthrene when added to the hexadecane caused increased emissions of these compounds as well as that of the nitro forms, but with no increase in black carbon emission. On the other hand, 1-methylnaphthalene, acenaphthene, and benzo(a)pyrene increased the emissions of some polynuclear aromatics such as fluoroanthene and altered the patterns of nitro-PAH emissions in hexadecane burning.

Furnaces

The strongly carcinogenic substance napthol (2,1-8-qra)naphthacene has been detected in charcoals issuing from public garbage furnaces and a domestic wood burner (Yasuhara et al., 1982). The concentration for the former is 14.9 ppm and for the latter 0.15 ppm. This compound has an extremely high activity to induce microsomal zosazolamine hydroxylase.

4

THE DEGRADATION OF BLACK CARBON

INTRODUCTION

There must be processes for the destruction of black carbon at the earth's surface. This can readily be seen by comparing the conservative estimate of yearly production from biomass burning, 0.1×10^{15} g (Seiler and Crutzen, 1982), with the total amount of carbon in the pools consisting of the biosphere, atmosphere, and oceans, 3×10^{18} g. Without a degradation mechanism, the earth's surface carbon would be converted to charcoal in a time period of $< 100{,}000$ yr.

Two mechanisms, photochemical and microbial breakdown, for carbon degradation have been proposed and both seem reasonable. Yet the evidence to support their occurrence in the environment from laboratory experiments is not especially convincing. Still, photochemical and microbial reactions with black carbon can provide environmental sinks.

Clearly, consideration can be given to a nonphotochemical inorganic degradation of environmental carbons, similar to that of the removal of carbon atoms from graphite by oxygen gas. A mechanism for this reaction has been proposed by Yang and Wong (1982). Although the experiments were carried out at 650°C, the results may be applicable at much lower temperatures. The rate of removal of carbon atoms at the graphite surfaces actually declined as the population of active sites for oxygen uptake increases. Further, the carbon removal continued even after the oxygen supply to the graphite was cut off.

Two processes were considered for the carbon removal by oxygen: (1) a direct reaction with oxygen in the gas phase and (2) one with the migrating oxygen atoms of the surface oxides. Only the second process is influenced by the population density of the active sites since the nonactive sites are shared as the collecting sites for the oxides. The migration of adsorbed oxides continues even after the oxygen supply has been cut off from the graphite and then the removal by the second mechanism can take place.

The oxidation of graphites by hypochlorite (and hypochlorous acid) has been studied at low temperatures 50–90°C (Ksenzhek and Solovei, 1960) and their results may be relevant to environmental black carbon degradation. The rate of oxidation increases sharply for pHs < 5.6:

$$2HClO + C = CO_2 + 2H^+ + 2Cl^-$$

$$2ClO^- + C = CO_2 + 2Cl^-$$

The reactions proceeded more rapidly in solutions containing excess sodium chloride. Clearly, conditions to match those in the laboratory may be found in nature. Perhaps, the production of chlorites photochemically may provide the oxidizing agents for the above reactions.

Some substances are resistant to microbial degradation. Alexander (1981) had considered the cases of recalcitrant synthetic organic molecules, situations that might well apply to that of black carbon. One explanation for the lack of, or slow, attachment by microorganisms upon a substance might be that it is removed form the "mainstream of catabolic pathways to be substrates for any species." The course of biological evolution may not have provided for the development of an enzyme to decompose the substance of concern, whether the proposed reaction is intra- or extracellular. Or, perhaps as Alexander (1981) points out, there may be an inaccessibility of specific sites on the black carbon (i.e., functional groups that can provide energy to the microorganisms at which the enzymes can function). It may be that black carbons do not provide an appropriate substrate.

A principal inadequacy with laboratory experiments so far conducted is the lack of knowledge of what is actually being degraded. Black carbon is an impure material containing hydrocarbons and other organic substances as well as the polymeric carbon moiety. Such impurities can be in the percent or more range. Thus, where a degradation experiment is carried out in the laboratory and only a few percent of the substance is destroyed, the question remains as to what is actually being decomposed—the carbon or some impurities?

On the other hand, there is evidence for the long time persistence of charcoals in some unconsolidated sediments. On the bases of the size distributions of charcoals in deep-sea sediments, Herring (1977) does not find any deterioration with age going back to 65 million years. He argues that any degradation will affect initially the smaller particles which have a higher surface area to mass ratio. He found no trend in size distributions in four stations in which he determined the diameters of about 10,000 particles to be between 0.6 to 32 μm. This can be seen in his analyses from Station 310, taken from a depth of 3516 m at 36° 51′ S and 176° 54′ E:

Age of Charcoal (10^6 yr)	Particle Diameter (μm) First Quartile	Median	Third Quartile
0.5	1.8	4.4	10
2.5	1.8	3.7	12
5.4	2.0	5.2	14
11	1.2	3.0	14
50	1.8	2.5	6
65	1.0	1.5	6.4

The ages above were determined by paleontological methods.

PHOTOCHEMICAL DEGRADATION

Elemental carbon in the presence of oxygen is thermodynamically unstable as can be seen from

$$C(s) + O_2\ (g) = CO_2\ (g)\ \Delta\ F°_{298°} = -94{,}052\ \text{cal/g}$$

The reaction is extremely slow under normal conditions and direct measurements of this reaction are yet to be carried out in the absence of catalysts.

The decomposition of charcoal, as well as a host of organic compounds and of diamond, in the presence of water, hydrogen ion, and air has been studied by Knoevenagel and Himmelreich (1976). Activated charcoal was dispersed in water in very small particles (sizes not given) and were kept in dispersion. Ultraviolet (UV) light irradiation (spectrum not defined) was applied. The results for the photoxidation of active charcoal, diamond dust, and hydroquinone were

Substance	Weight (mg)	HCl (0.1 *N* in ml)	Time (hr) for the Formation of CO_2 (% of Theoretical)		
			25%	50%	75%
Activated charcoal	120	34.5	135.5	282.3	423.0
Diamond dust	24	0	16.4	42.4	71.3
Diamond dust	24	69.0	19.6	69.8	121.3
Hydroquinone	110	0	10.3	22.9	43.7
Hydroquinone	110	69.0	6.3	17.2	32.0

The temperature of the reactions was 90–95°C.

The results are difficult to assess and this type of experiment should be repeated under more rigorously controlled conditions. It is not possible to ascertain whether the diamond and charcoal values were determined throughout the entire time interval or represent extrapolations. The authors state "These degradation curves for nearly all of the compounds were documented to the end." The possibility that the charcoal and diamond curves are extrapolations exists. The greater resistance of charcoal than diamond dust to photochemical oxidation is difficult to understand. Still, the results, if valid, do bear upon the environmental fates of charcoals. Even the high laboratory temperatures of 90–95°C do not detract from the strong possibility that there is a photochemical sink of the large amounts of charcoals formed in history and prehistory.

MICROBIAL DEGRADATION

Two investigations, undertaken over an interval of >50 yr, indicate that microorganisms can oxidize charcoals, coals, and peats. The first (Potter, 1908) involved simple, but elegant, experiments to show that carbon dioxide is released following exposure of amorphous carbons to bacteria isolated from soils. The second (Shneour, 1966) utilzed C-14, incorporated into elemental carbon, to establish a microbially mediated oxidation to carbon dioxide.

Potter (1908) employed three techniques to establish his argument. The initial strategy used the inoculation of the wood charcoals with bacteria and the passage of an airstream through an incubator. Any carbon dioxide produced was removed from the airstream by sorption into a barium hydroxide solution. A subsequent titration with standard oxalic and hydrochloric acids indicated the amount of carbon that was oxidized. The charcoals were initially freed of organic matter by heating to 1200°C in a crucible protected from access to atmospheric oxygen. The apparatus for the inoculated and the control samples of carbon was extensively cleaned with boiling nitric acid to eliminate any possible organic contamination. The bacteria inoculum appeared to consist of a pure culture of *Diplococcus sp*.

Carbon dioxide could not be detected in the airstreams issuing from either the uninoculated control or inoculated flasks during the first week of the experiment. However, at the end of the second week, the latter did produce measureable quantities of carbon dioxide while the controls exhibited no traces of the gas. The lag time, Potter (1908) explained, was probably due to the initial uptake of carbon dioxide by calcium oxides produced during the heat treatment of the charcoals. This hypothesis was strengthened with the observation that carbon dioxide was removed from airs by freshly heated charcoals.

Even after periods of a month, the carbon dioxide evolution from 5 g of charcoal did not exceed 8 mg/week, for example, under a percent of the weight of the carbon was oxidized. Whether this reacted material was carbon or some impurity in the charcoal cannot be ascertained today.

In order to eliminate the possibility that carbon dioxide came from the bacteria, bacteria were introduced to distilled waters in the absence of any charcoal. No carbon dioxide was evolved. This study constituted his second approach to the problem.

A third attack determined the temperature rise in an incubator containing the charcoal and the bacteria in relation to a control to which no bacteria were added. Since the oxidation of charcoal by oxygen is exothermic, any temperature rise in the incubator would be indicative that a reaction is occurring. Again, precautions were taken to insure that the apparatus was free of organic matter. Whereas the sterile flask indicated no temperature increase, the inoculated one showed a temperature rise after two days. A maximum elevation of 0.19°C was noted in the inoculated flask after a period of 6 days. The temperature was maintained for another week before the apparatus was disassembled. These experiments led Potter (1908) to submit that there can be a slow oxidation of amorphous carbons through the agency of microorganisms.

Shneour (1966) introduced carbons, produced by the reduction of $C^{14}O_2$ with magnesium, to two soils from forest fire areas and to one from an urban site. The indigenous miroorganisms of the soil are presumed to be the active agents in the oxidation of black carbon. Control soil samples were sterilized by their exposure to: temperatures of 200°C for 72 hr, steam at 135°C for 1 hr, an electron beam of 5 Mrad, chloroform for 24 hr, or toluene for 24 hr. All techniques appeared to have essentially the same

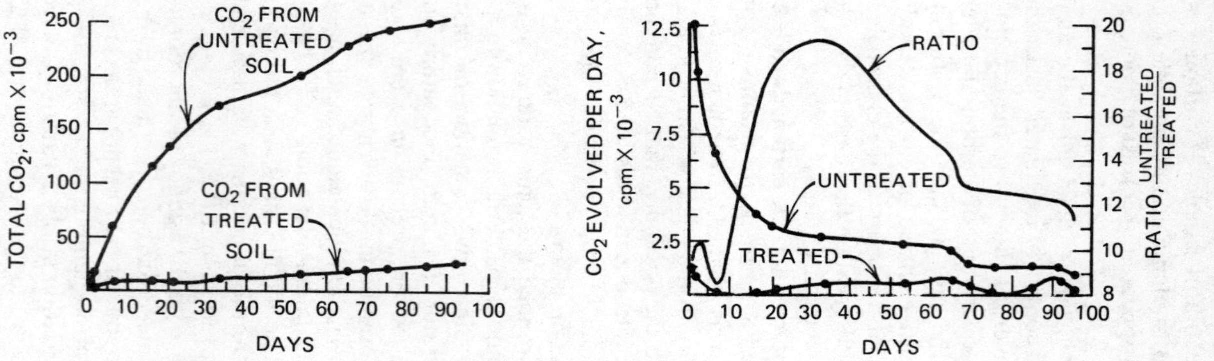

Figure 4.1. Total CO_2 evolved by an urban soil, treated and untreated (left). Specific activity of $^{14}CO_2$ evolved by an urban soil, treated and untreated (right). (From Schneour, 1966). Reproduced with permission of the author and the American Association for the Advancement of Science.

effect. Dry heating at 200°C for 72 hr in an argon atmosphere was used for most of the experiments.

After 96 days, the total radioactivity in the carbon introduced to the soil from the urban site was reduced by about 2% (Figure 4.1). The treated control produced $< 10\%$ of the $C^{14}O_2$ as did the untreated sample. From this, it can be concluded that microorganisms can degrade elemental carbon or the organic phases associated with the elemental carbon.

There was a progressive decrease in the specific activity of the C-14 in the untreated urban soil indicating that the artificially produced carbons were more readily degraded than were naturally occurring black carbons (Figure 4.1).

5

ANTHROPOGENIC BLACK CARBONS

INTRODUCTION

Human society inadvertently and deliberately introduces black carbons to the environment. The intentional burning of forests and fossil fuels produces black carbon as a consequence of the incomplete combustion of organic materials. On the other hand, there is a variety of impure elemental carbons produced commercially that enter our surroundings, usually unintentionally. The annual production of the commercial black carbons [$(4–5) \times 10^{12}$ g/yr] is dwarfed by the estimated influx of charcoals from biomass burning [$(0.5–2) \times 10^{15}$ g/yr] to the environment (Seiler and Crutzen, 1980).

In some countries of the world, especially those not as yet industrialized, charcoal can be the main domestic fuel (Foley and van Buren, 1982), however, the consumption and production on a worldwide basis has not as yet been made. A sense of the importance of this fuel may be seen by consideration of its use in a country where population is slightly over 0.1% of the world's total. In Senegal, whose citizenry numbers around 5.7 million, there is an estimated usage of charcoal of 2.5×10^{11} g/yr (Foley and van Buren, 1982). The amount of this charcoal that goes to the environment unburned is impossible to estimate. There are no attempts made to burn the charcoal powders that come about from the handling and transport of the material. Further, there is an accumulation of unburned charcoals around the primitive kilns.

In this chapter we will review the types of black carbons made commercially and indicate their characteristics. Much of the information comes from a survey by Fitch and Smith (1979). There is emerging a clear picture of the associated organic and inorganic substances as a function of the charcoal source.

TYPES OF ANTHROPOGENIC BLACK CARBONS

Graphite

Most of the commercial graphite is prepared by the high temperature conversion of impure carbon blacks or coke. (Clearly, there may be a semantic problem in calling commercially produced graphite a black carbon. Still, it is an indirectly made product of biomass combustion.) High temperatures are required for graphite formation and as a consequence few small organic molecules survive. For example, Fitch and Smith (1979) found only trace quantities of naphthalene in the sample they analyzed.

Active Carbon

Wood, coal, or petroleum residues are pyrolized and oxidized under controlled conditions to produce black carbons with highly porous structures. They are used extensively in water purification programs whose demands for this product will most probably increase in the future. The black carbons have extremely small amounts of extractable organic adsorbates. Fitch and Smith (1979) found only trace quantities of low-molecular-weight polynuclear aromatic hydrocarbons in one of the two samples analyzed.

Carbon Blacks

The annual world production of carbon black in the early 1970s was about 4–5 million tons. The most important use is in compounding rubber for automobile tires. Here, it not only acts as a pigment, but also as a filler to improve the tire quality by reducing its resistance to abrasion and hence wear through as immobilization of the rubber (Medalia, 1974). About 90% of the carbon production in the United States is used in tires. It is also used in printing inks. Because of difficulties in removing it from paper, it impairs the recycling process.

Carbon blacks are produced by the reactions or hydrocarbon liquids or gases with a limited supply of air at temperatures between 1320 and 1540°C. The unburned carbons range in size from 10 to 500 nm, as fluffy, often spherical particles. They are primarily aciniform carbon (Medalia and Rivin, 1982).

In the channel or contact process, smoky flames from tiny jets impinge upon iron channels from which the deposited carbon particles are removed by moving the channels over stationary scrapers. The fuel is natural gas or petroleum. The size range of the particles is from 10 to 30 nm. The oxygen contents can achieve levels up to 20%. Channel blacks are the most expensive of the carbon blacks and are used as fine pigments.

Lampblack was the classic pigment used in printing and has been replaced to a large degree by the furnace and channel blacks. It is produced by the incomplete combustion of heavy petroleum fractions. It is composed of big particles, 40–200 nm and normally contains large amounts of polynuclear aromatics.

In the thermal process, hydrocarbon gases are decomposed to hydrogen and carbon through contact with a heated refractory. The process is cyclic and begins with the heating of a brick checkerwork to about 1300°C by a mixture of air and fuel gas. The heating is stopped by cutting off the fuel supply and natural gas is introduced into the furnace where it is decomposed by the heat from the bricks. The effluent gases, containing the thermal black particles and the unreacted natural gas, are cooled by a water spray to 125°C and passed through cyclonic collectors and fabric filters to recover the carbon black. The effluent gases are cooled and compressed and then passed through the furnaces as fuel. When acetylene is used as a feed, the acetylene black produced is characterized by an extremely high electrical conductivity. It is used in the manufacture of dry cells.

The oil furnace process utilizes liquid hydrocarbon as feedstock that is fed into the combustion zone of a natural gas-fired furnace. Quench waters cool the gases to 540°C to stop the cracking. The gases carrying the carbon particles are further cooled to 450°C through passage into heat exchangers and by direct water sprays. The products are transported to cyclones and bag filters. Particle sizes range principally from 20 to 50 nm. The particles contain polynuclear aromatic hydrocarbons as well as sulfur and oxygen containing compounds.

In the United States the oil furnace process accounts for 90% of the production, the thermal process for about 10%. Lampblack and acetylene black are produced at only two plants and account for $<1\%$ of the carbon black output. The gas furnace process is being phased out and the last channel black plant was closed in 1976 (EPA, 1979).

Before 1940, channel black constituted about 90% of the total carbon black production, but by the middle 1960s, this had dropped to $<6\%$. The oil furnace process replaced the channel process as a consequence of yielding a better product and of the increasing price of natural gas relative to petroleum during this period.

Fitch and Smith (1979) found phenanthrene in an extract from one commercial variety of channel black, the only polynuclear aromatic detected in the two analyzed samples. However, a variety of oxidized species were observed such as 9-fluorenone, anthraquinone, 1-nitronaphthalene, 1,8-naphthalenedicarboxylic anhydride, and dimethyl phthalate.

The morphologies of the small size fraction of carbon blacks may be characteristic of their mode of formation (Medalia and Heckman, 1969). The number of particles per aggregate (see Figure 3.1) are highest for acetylene black, low for thermal black, and intermediate for furnace black. Perhaps, such measurements, if extended to other black carbons in the environment, might prove useful in identifying their sources.

Forest Fire Burning

The flux of particulates from forest burns may be of greater significance with respect to air quality than those of gaseous products (Murphy et al., 1970). The solid particles, including the charcoal, reduce visibility through their absorption and scattering of sunlight. These authors point out that whereas prescribed burning produces <2% of the atmospheric burden of particles, wildfires may introduce possibly five times as much. A variety of strategies are proposed to reduce the impairment of air quality. Many meteorological tactics can be effective. For example, controlled burning during midday takes advantage of convective heating of the atmosphere. Burning during those days when the air is moist provides the possibility that the air will become involved in active condensation subsequently, a process that will remove the particles from the air. Research into burning methods with electrical ignition and fuel boosters creates hotter fires of shorter duration. The use of napalm grenades has been especially effective along these lines. Burning when meteorological conditions are such that the particles will not be carried into urban or recreational areas is clearly desireable. Finally, alternate disposal techniques to burning, such as crushing of slash are worthy of assessment.

The particulate emissions from a forest burn are dependent upon the nature of the fuel and upon the intensity of the burn, measured in British thermal units per square foot per minute (Btu/ft^2/min), according to Sandberg (1974). The combustions took place in the laboratory and involved fine fuels, <3 in. in diameter:

Fuel	Emissions in (lb/ton) of Fuel Burned
Ponderosa pine	12.5
Douglas fir without needles	6
Douglas fir with needles	24

In addition there was an approximate inverse proportionality between fire intensity and emission factors. When the fire intensity doubled, the emission factor was reduced about 45%.

There are several implications from this work. First of all, in controlled burning the duration of the combustion should be minimized. Secondly, the high water content needles increase particulate flux and their burning should be avoided when possible. In summary, the ideal fire with respect to minimum emissions involves a fast burn that is hot, with low contents of high water content substances.

The effective use of charcoal concentrations in sediments to measure the extent of biomass burning may be jeopardized by the extensive use of fire retardants. Sandberg et al. (1975) noted an increase in the emission factors for CO, hydrocarbon gases, and particulate matter when ponderosa pines were burned after treatment with diammonium phosphate flame retardant in comparison with the emissions from untreated fuel. Their results may be summarized in the following way:

	Emissions (lb/ton) of Initial Fuel Bed Weight			
	Untreated Fuel	Treated Fuel (gal/100 ft^2) 3	6	12
Particulate matter	9 ± 1	19 ± 4	20 ± 3	19 ± 3
Hydrocarbon gases	8 ± 2	11 ± 4	12 ± 2	12 ± 4
Carbon monoxide	146 ± 10	174 ± 17	173 ± 12	143 ± 56

It is interesting to note that there are only significant differences between the untreated fuels and the treated fuels. Irregardless of the amount of treatment, the emissions are about the same. If the particulate matter is primarily composed of charcoal, then this type of treatment can alter the true extent of biomass burning.

Wood Burning

Residential firewood use is concentrated in the urbanized areas of the northeast and north central states of the United States and accounts for 9–11% of U.S. space heating (Lipfert and Dungan, 1983).

In some cities in the United States, wood burning can account for a

Table 5.1. Carbon Emissions from the Combustion of Woods in Domestic Fireplaces (Muhlbaier and Williams, 1982)[a]

Type	Organic C	Elemental C g/kg wood	Remainder
Softwood	2.8 ± 1.5 (45%)	1.3 ± 0.5 (21%)	2.1 ± 0.6 (34%)
Hardwood	4.7 ± 3.4 (47%)	0.39 ± 0.34 (4%)	4.9 ± 4.2 (49%)
Synthetic log	1.7 ± 1.1 (24%)	3.5 ± 3.3 (49%)	2.0 ± 1.2 (28%)

[a] Reproduced with permission of the authors and Plenum Press.

substantial amount of the particulates in the winter atmospheres. For example, in the Denver, Colorado area, about 20–30% of the aerosols are attributed to this activity (Dasch, 1982) and 50% in Portland, Oregon. In the former case, the charcoal contents of the particulates were measured as a function of both the type of wood and that of the combustion chamber. Softwoods averaged 33 ± 13% charcoal whereas hardwoods averaged 8 ± 7%. The investigator took special pains to ensure that the reported values of elemental carbon did not contain contributions from charred materials. The fireplace burning season appears to cover about 150 days in the Denver area. The carbon emissions from domestic fireplaces have been studied by Muhlbaier and Williams (1982) (Table 5.1). Both hardwoods and softwoods were combusted. The investigators attribute the higher production of charcoal from softwoods as a consequence of their high content of lignin that previously had been shown to promote charcoal formation.

Muhlbaier and Williams (1982) have estimated the annual emissions from gas furnaces, fireplaces, and automobiles in the United States for 1980 (Table 5.2). Fireplaces are about an order of magnitude greater in the production of black carbons than are automobiles or residential gas furnaces.

Residential wood combustion has been shown to contribute about 75% of the black carbon in aerosols from Elverum, Norway, although this burning only accounts for 3% of the energy production (Ramdahl et al., 1984). This conclusion was based upon the C-14 content of the black carbon with the assumption that wood burning produced a specific C-14 activity of modern day CO_2 and that fossil fuel burning produced C-14 specific activities of zero. The town, situated about 120-km north of Oslo has some light industries with no significant process emissions.

Table 5.2. Estimated Annual U.S. Emissions of Particulate Carbon for 1980 (Muhlbaier and Williams, 1982)[a,b]

Source	Organic Carbon	Black Carbon
Residential gas furnaces	0.037	0.018
Fireplaces	86	11
Automobiles		
Precatalyst	5.6	2.0
Catalyst	2.0	1.5
Diesel	1.1	4.2

[a] In units of 10^9 g/yr.
[b] Reproduced with permission of the authors and Plenum Press.

Automobiles

Automobile emissions were tested at sea level (Warren, Michigan) and high altitude (Denver, Colorado) by Muhlbaier and Williams (1982) (Table 5.3). Three driving conditions were used: Federal Test Procedure (FTP) urban cycle, and Highway Fuel Economy Test (HFET) cycle, and constant speed movements. The reduced air pressure in Denver has a direct effect on the air/fuel ratio, which governs the amounts of gaseous and particulate emissions. Soot emissions increased 290 and 600% for the precatalyst and catalyst car, respectively, and decreased slightly for the diesel.

The oxidation catalyst equipped automobiles had significantly smaller black carbon and organic carbon emissions. This results not only from the effect of the catalyst but from other variables such as fuel composition and other emission controls.

Vehicular emissions collected in Pennsylvania Turnpike tunnels contained about 68 ± 11% carbon by weight (Pierson, 1979) of which about 40% is composed of alkanes extractable into benzene or *ortho*-dichlorobenzene. Aromatics and olefins were essentially absent. About 50% of the carbon is not extractable and is assumed to be elemental.

Warner (1976) indicates that the submicron carbon black particles, abraded from automboile tires during their normal wear, may have long persistances in the atmosphere and by inference in the environment. The carbon particles associated with the butadiene–styrene polymer system (and perhaps other tire components) are stripped from the tire surface. It may be that these degradation products can be characterized by their

Table 5.3. Summary of Carbon Emission from Automobiles (Muhlbaier and Williams, 1982)[a]

Car Type	Driving Mode (see text)	Total Particulate	Organic Carbon (mg/mile)	Black Carbon
		Warren, Michigan (180 m above sea level)		
Diesel	FTP	680	137	520
	88 km/h	424	157	258
Precatalyst	FTP	119	10	3.6
	HFET	196	9.3	1.9
Oxidation	FTP	14	3.0	2.5
Catalyst	HFET	25	4.2	1.5
		Denver, Colorado (1550 m above sea level)		
Precatalyst	FTP	152	43.8	24.1
	HFET	300	46.2	20.9
Oxidation	FTP	18.6	4.9	5.6
Catalyst	HFET	29.5	8.9	3.1

[a]Reproduced with permission of the authors and Plenum Press.

morphologies and shapes and can be readily identified in areas adjacent to well used highways.

The component of the Denver aerosol responsible in large part for light absorption is perhaps abraded tire rubber containing black carbon (Patterson, 1979). The significant property of the tire rubbers is their large size, compared to the smaller soot particles.

COMPARISON OF PROPERTIES OF ANTHROPOGENIC BLACK CARBONS

Commercial Products

For various anthropogenic black carbons Fitch and Smith (1979) have measured variations in composition, size, amount of extractable organic matter, and their ability to adsorb benzo(a)pyrene from solution (Tables 5.4 and 5.5). The coarsest material was graphite and the finest, some of the channel blacks. The lampblack had the highest carbon content

Table 5.4. Particle Sizes of Some Anthropogenic Black Carbons (Fitch and Smith 1979)

Name	Source	Trade Name	Particle Size (nm)
Channel 19	*b*	Colour Black FW2000	13
Channel 18	*c*	Neo Spectra AG	13
Channel 15	*b*	Colour Black FW2	13
Channel 13	*b*	Special Black 5	20
Channel 12	*b*	Special Black 4	
Channel 10	*c*	Neo Spectra Mark III	14
Channel 4	*b*	Colour Black S160	20
Active 9	*a*	Darco G60	
Active 8	*d*	Norit A	
Furnace	*b*	Corax L	23
Lampblack	*a*	Lampblack	44
Graphite	*a*	Power Grade 38	<44,000

[a]Fisher Scientific Company.
[b]Degussa Inc., Pigments Division.
[c]Cities Service Company, Columbian Division.
[d]Matheson Coleman and Bell.
Reprinted with permission of the American Chemical Society.

Table 5.5. Analytical Data on Anthropogenic Black Carbons (Fitch and Smith 1979)

	Elemental Analysis							
Name	C	H	N	S	O	Ash	Extract[a]	Adsorptivity[b]
Channel 18	81.3	0.5	0.0	0.1	18.0	0.1	<0.1	0.030
Channel 19	79.2	0.7	0.4	0.4	19.4	0.0	1.4	0.056
Active 8	88.0	0.6	0.1	0.0	7.6	3.7	<0.1	0.060
Channel 15	83.9	0.3	0.4	0.3	14.9	0.1	1.1	0.069
Channel 10	89.5	0.7	0.0	0.0	9.8	0.0	<0.1	0.072
Active 9	88.7	0.7	0.0	0.0	8.8	1.8	<0.1	0.083
Graphite	97.0	0.1	0.0	0.0	1.6	1.3	<0.1	0.14
Channel 13	85.6	0.6	0.3	0.4	13.2	0.0	0.85	0.15
Furnace	96.9	0.3	0.0	0.7	2.1	0.0	<0.1	0.23
Channel 12	86.0	0.7	0.4	0.4	12.4	0.0	0.48	0.36
Channel 4	94.6	0.6	0.1	0.3	4.4	0.0	1.0	3.8
Lampblack	96.7	0.6	0.0	1.5	0.9	0.3	0.37	40.0

[a]Weight percentage of the benzene/methanol extract.
[b]Distribution coefficient for 10 ng of benzo(a)pyrene between toluene (2 mL) and carbon (10 mg). Results in units of nanograms per milliliter of toluene per nanograms per milligram of carbon (Fitch and Smith, 1979).
Reprinted with permission of the American Chemical Society.

(96.7%) whereas the lowest carbon content was found in a channel black with a value slightly under 80%.

In their samples (Tables 5.4 and 5.5), Fitch and Smith (1979) observed that particle size and the consequential surface area are not the only factors regulating adsorption. For example, graphite with a small surface area per unit weight has an intermediate adsorptivity.

Most of the uses of carbon blacks are dispersive and one can expect a large percentage of the 4 or 5 Mton annually produced to end up in the environment. The particulate emissions from both the oil furnace and thermal processes are primarily carbon. EPA (1979) indicated that in the latter process the particulate emissions are negligible, although emission data are not available. On the other hand, in the oil furnace process, there are measurements of emissions and the primary output is from the main process vent where fluxes on the order of several kilograms per ton of carbon produced are observed (EPA, 1979).

EMISSION CHARACTERISTICS

The environmental behavior of black carbons introduced to the atmosphere from anthropogenic combustion processes depends upon the characteristics of the source, aerosol properties, its chemical composition, and meteorology (Charlson and Ogren, 1982). The amount of black carbon produced per amount of fuel burned depends primarily upon the type of burning as well as upon the specific device. Table 5.6 gives emission factors for a variety of processes. Perhaps more important

Table 5.6. Emission Factors of Black Carbons from Different Combustion Processes (Adapted from Charlson and Ogren, 1982)[a]

Fuel	Source	(grams black carbon/kg)
Natural gas	Steam generator	3×10^{-4}
	Domestic water heater	0.1
	Heating boiler	0.01–0.07
Gasoline	Automobile engine	0.1
Diesel	Automobile engine	2–4
	Truck/bus engine	0.6–1
Jet A	Aircraft turbine	0.5–3
Fuel Oil Number 2	Utility turbine, 20 MW_e	0.08

[a]Reproduced with permission of the authors and Plenum Press.

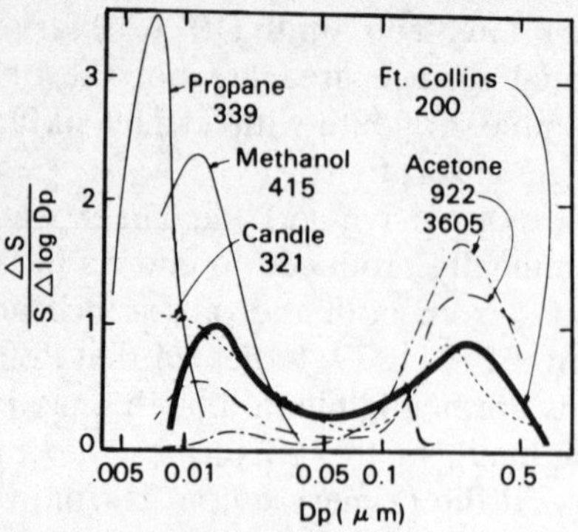

a. Laboratory aerosols [9].

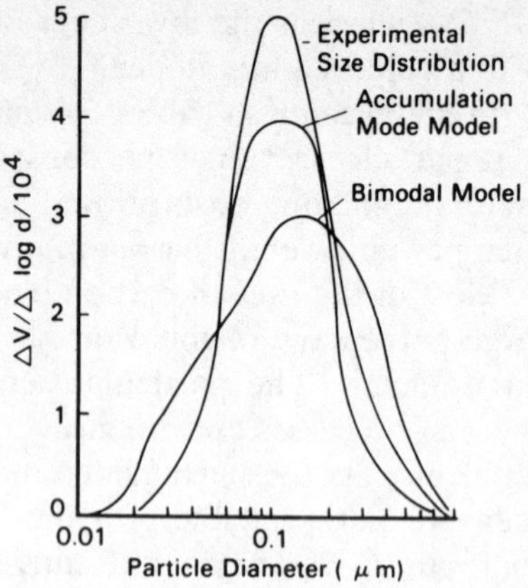

b. Diesel aerosol [10].

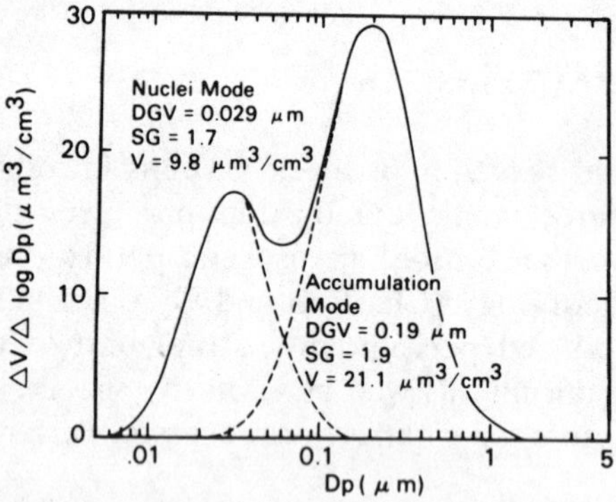

c. Freeway aerosol [9].

Figure 5.1. Size distributions of black carbons from (*a*) Laboratory aerosols, (*b*) diesel emissions, and (*c*) freeway aerosols (Charlson and Ogren, 1982). Reproduced with permission of the authors and Plenum Press.

is the size distribution of the emitted particles. Figure 5.1 provides measured and theoretical size distributions of laboratory and field-collected black carbons. The chemical composition may be important in governing whether or not the black carbon is hygroscopic and can be involved in the oxidation of sulfur dioxide to sulfuric acid.

Once the black carbon enters the atmosphere, its nature can be altered by processes occurring there. Charlson and Ogren (1982) indicate that these combustion particles have a strong tendency to coagulate. As the particles increase in size, dry-fallout processes are enhanced. The atmospheric lifetimes of these small particles are determined by the concentrations of the introduced particles and those of the preexisting fine particles. In urban areas, the lifetime appears to be of the order of 1 hr, while in the relatively unpolluted upper troposphere it might approach times as long as days.

The chemical makeup of the black carbons regulate absorption processes of other atmospheric materials. However, the general inertness of black carbons indicates that removal processes are primarily physical.

After injection into the atmosphere, the black carbon particles are subjected to the vagaries of wind systems. Charlson and Ogren (1982) argue that there should be a background level of black carbon, as a result of the residence times of days and the large number of geographically separated sources.

SOURCE FUNCTIONS

A mass balance sheet has been constructed for the emissions of black carbon to the Los Angeles atmosphere during January and February, 1980, by Cass et al. (1982). Forty-six metric tons of aerosol carbon are emitted daily of which about one third is characterized as elemental carbon (Table 5.7). The elemental carbon is primarily in particles $<10\ \mu m$.

Table 5.7. Emissions Estimates (Cass et al., 1982)[a]

	Estimated 1980 Fuel Use (10^9 Btu/day)	Fine Nonvolatile Carbon (kg/day)	Total Aerosol Carbon (kg/day)
	Mobile Sources		
Highway vehicles			
Catalyst autos lt. trucks	963.78	369	803
Noncatalyst autos lt. trucks	566.77	563	7042
Medium heavy gasoline vehicles	249.26	286	3571
Diesel vehicles	137.16	4319	5917
LPG[b] for carburation	6.07		
Civil aviation			
Jet aircraft	45.50	547	749
Aviation gasoline	1.32	1	16

Table 5.7.(*Continued*)

	Estimated 1980 Fuel Use (10^9 Btu/day)	Fine Nonvolatile Carbon (kg/day)	Total Aerosol Carbon (kg/day)
Commercial shipping			
Residual oil-fired ships boilers	32.05	67	333
Diesel ships	22.43	540	740
Railroad			
Diesel oil	38.11	1530	2096
Military			
Gasoline	6.03	7	86
Diesel oil	17.81	686	940
Jet fuel	16.71	462	633
Residual oil (bunker fuel)	0.27	1	3
Miscellaneous			
Off-highway vehicles (diesel)	39.73	1531	2097
Total		10,909	25,026
Stationary Sources			
Fuel combustion			
Electric utilities			
Natural gas	666.25	...	48
Residual oil (0.25% S)	885.72	312	1560
Landfill digester gas	0.85	...	0.06
Refinery fuel			
Natural gas	148.54	...	90
Refinery gas	347.63	...	210
Residual oil	5.57	3	15
Nonrefinery industrial fuel			
Natural gas	421.64	...	212
LPG	2.74	...	1
Residual oil	53.42	28	141
Distillate oil	42.74	85	147
Digester gas (1C Carbon Engines)	6.30	6	27
Coke oven gas	37.53	...	19
Residential/commercial			
Natural gas	1592.26	483	1465
LPG	18.08	6	17
Residual oil	22.19	12	59
Distillate oil	22.19	41	70
Coal	0.55	27	124
Total		1003	4205

	Fine Nonvolatile Carbon (kg/day)	Total Aerosol Carbon (kg/day)
Industrial Process Sources		
Industrial process point sources		
Petroleum industry		
Production	0	≤103
Refining	7	30
Marketing	0	<207
Organic solvent use		
Surface coating	0	1511
Degreasing	0	≤21
Other	0	≤10
Chemical		399
Metallurgical	111	3686
Mineral	...	...
Waste burning at point sources	37	≤56
Wood processing	0	≤79
Food and agriculture	0	275
Miscellaneous industrial	0	164
Total	155	≤6541
Fugitive Sources		
Road and building construction	...	...
Agricultural tilling	...	...
Refuse disposal sites	...	...
Livestock feed lots	0	65
Unpaved road travel	...	...
Paved road travel	...	...
Forest fires (seasonal)	226	3761
Structural fires	73	149
Fireplaces	183	373
Cigarettes	17	1692
Agricultural burning	58	647
Tire attrition	1729	5240
Brake lining attrition	388	2159
Sea salt	...	...
Total—Annual average day	2774	14,086
Total—Winter day with no forest fires	2448	10,325

[a]Reproduced with permission of the authors and Plenum Press.
[b]LPG stands for liquefied petroleum gas.

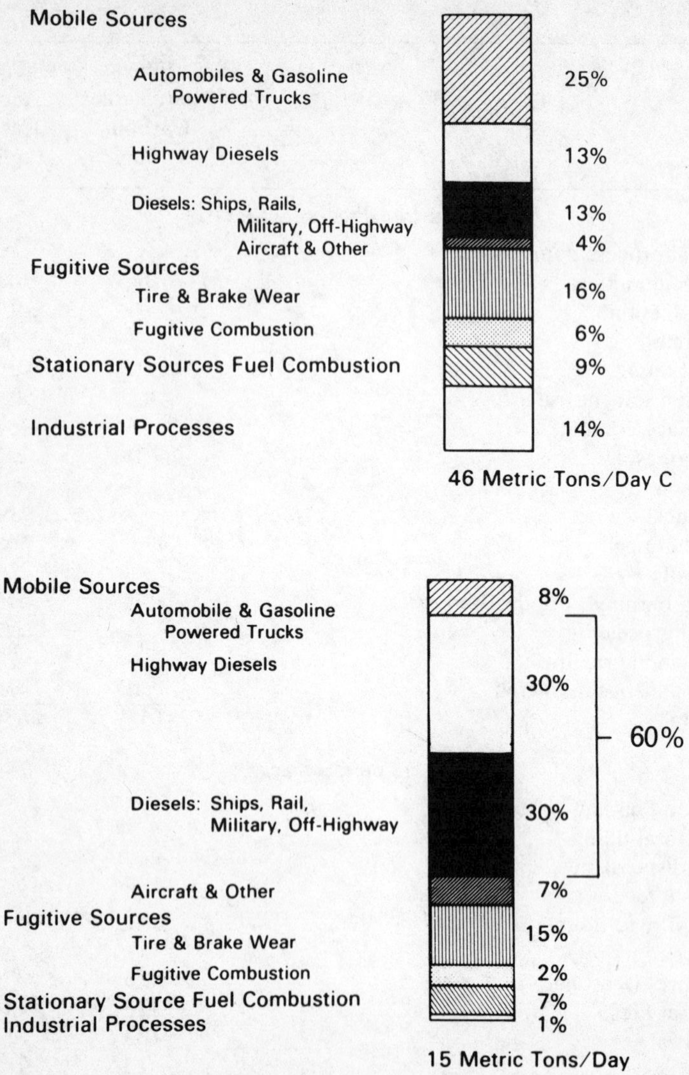

Figure 5.2. Sources of black carbon in greater Los Angeles atmospheres, January–February 1980. (Upper) total carbon-containing aerosols. (Lower) Black carbon. (Cass et al., 1982). Reproduced with permission of the authors and Plenum Press.

The major part of the emissions come from internal combustion engines burning light and middle distillate fuel oils (diesels and jet aircraft) (Figure 5.2). Diesel engines are a prime contributor. Although they burn <5% of the total fossil fuel used in the Los Angeles basin, they introduce about 60% of the black carbon emitted to the atmosphere. The total carbon to elemental carbon ratio is 3.2:1 for integrated highway emissions (vehicular, tire, and brake dusts). It is to be emphasized that the data, although given sometimes to more than five significant figures, are somewhat uncertain. Some of the numbers were derived from single experiments or were combined from separate investigations.

In Vienna, 80% of the black carbon in aerosols is attributed to diesel fuel engines on the basis of C/Pb ratios emitted from the exhausts and those in the aerosols themselves (Puxbaum and Bauman, 1984). The soils adjacent to highways might be expected to show elevated black carbon levels compared to those at greater distances. In one studied in my laboratory (Highway *I*-5, San Diego, California) the soils were not evidently influenced strongly by a vehicular source of black carbon:

Site (m east)	Black Carbon Concentration in Soil (wt%)
3	0.089
25	0.075
100	0.027
6400	0.080

6

BLACK CARBONS IN THE ENVIRONMENT

INTRODUCTION

There is a ubiquity about elemental carbon in the earth's environment. It is found in igneous, metamorphic, and sedimentary rocks. It is observed in the atmosphere and in rain. Further, lunar and meteoritic samples contain forms of elemental carbon. There are probably historical records of atmospheric fallout of black carbon in glaciers as there are in lacustrine and marine sediments. Some occurrences of graphite in the environment are also considered here. This mineral can enter the major weathering cycle and become associated with black carbons as we have defined them. Finally, black carbon is found in human lungs from the inhalation of aerosols. Each environmental carbon has a story to tell.

OCCURRENCES OF ELEMENTAL CARBONS

Moon

There is a suggestion that elemental carbon exists on the Moon's surface (Chang et al., 1970). Lunar fine particles from Mare Tranquillitatis were subjected to pyrolysis at 750 and 1050°C. In the former treatment, an organic residue survived, however, when the residue was heated to 1059°C, it was combusted totally to carbon dioxide. The investigators suggested that some of the residual carbon could have been in the form of elemental carbon.

Meteorites

Elemental carbon has been identified as a constituent of carbonaceous chondritic meteorites. Unusual interest has been directed towards it inasmuch as the carbon phases carry high amounts of noble gases, whose origin is not as yet established. These gases have an unusual composition, being depleted in the lighter members. Origins have been associated with the fissioning superheavy elements or presolar stellar sources. There appear to be at least four carriers of the gases: chromite and three materials containing carbon. A number of laboratories have examined isolates of the carbon and have described it as amorphous on the basis of X-ray diffraction, electron diffraction, and electron microscope studies. Recent investigations claim that the carbon is in the form of carbynes, a series of carbon polymorphs containing triply bound carbons. These materials constitute a low pressure form of carbon, stable above 2600°C. A review of the literature on the subject is provided by Smith and Buseck (1981, 1982).

Smith and Buseck (1981, 1982) studied the carbon isolated from carbonaceous chondrites using high resolution transmission electron microscopy (HRTEM). The ability of the instruments to resolve the 3.4 Å spacing of the basal planes allows studies of the ordering of partially graphitized carbon.

The carbon particles through high resolution images were shown to consist of a tangled aggregate of somewhat fibrous crystallites with a prominent lattice fringe spacing of 3.4–3.9 Å. This suggests that the material is a poorly ordered graphite, where the lattice fringes correspond to the basal planes (002) of the graphitic structure. The lengths of the lattice fringes appear to be no more than 150 Å, but they are probably somewhat longer inasmuch as they can only be traced when the graphitic basal planes are almost exactly parallel to the microscope beam axis. A few crystallites were found to be 1000 Å in longest dimension.

Smith and Buseck (1981, 1982) were unable to find any carbyne grains by this technique. The structure they found is similar to that proposed by Ban et al. (1975) for carbonized polyvinylidene chloride (Figure 6.1) or to the glassy carbon described by Jenkins et al. (1972). The latter group made their carbon by the direct pyrolysis of a phenolic resin. The structure consists of intertwined crystallites comprising ribbon-shaped aggregates of graphitic basal planes. The crystallites are interlinked to form a complex three-dimensional structure. Clearly of the interest is the ability of this structure to sorb large amounts of gases. The "carbyne particle" identified previously by other workers was shown by Smith and Buseck (1982) to be a sheet silicate.

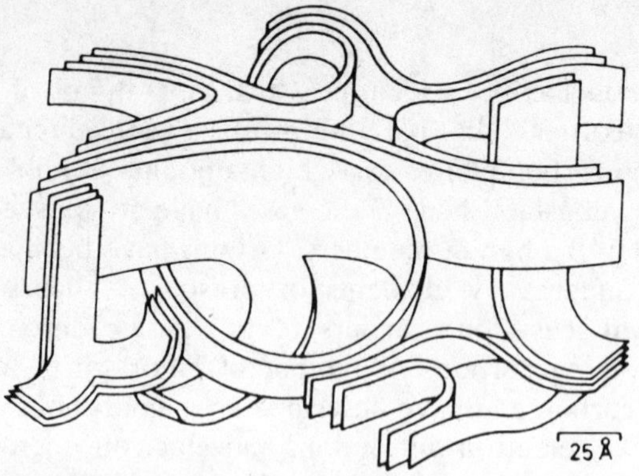

Figure 6.1. Proposed structure for elemental carbon from Allende meteorite (Smith and Busek, 1981a). Reproduced with permission of the authors and the American Association for the Advancement of Science.

The degree of ordering in a graphitic carbon of the Tieschitz chondrite was used to determine the maximum temperature to which the meteorite was heated after accretion (Michel-Levy and Lautie, 1981). Microscopic examination with a Raman laser microprobe characterized the isolated charcoal and show it to be a highly disordered graphite that upon heating became more ordered. However, the ordering only increased after heating

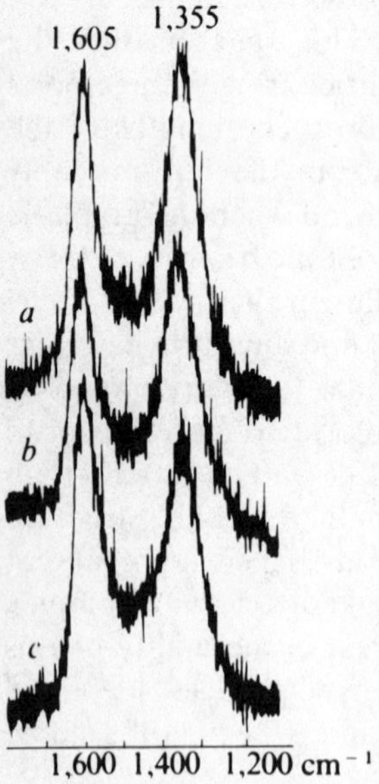

Figure 6.2. Raman spectra between 1600 and 1200 cm^{-1} for the Tieschitz meteorite carbon. Unheated sample (*a*), sample heated to 300° (*b*), and sample heated to 600°C (*c*), (Michel-Levy and Lautie, 1981). Reproduced with permission of the authors and *Nature*.

above 300 to 350°C, which placed a limit on the maximum temperature previously experienced by the object (Figure 6.2).

Submarine Basalts and Peridotite Nodules

The origin of elemental carbon in submarine basaltic glasses and mantle-derived peridotite nodules from alkali basalts is attributed to the disproportionation of CO by the reaction $2CO \rightarrow C + CO_2$ (Mathez and Delaney, 1981). The associated sulfide/oxide surfaces are presumed to catalyze the reaction initially; after the first deposition of carbon, the reaction becomes autocatalytic. Carbon was identified in five submarine basalts and two nodules by electron microprobe techniques. The high X-ray intensities indicated the form of the carbon to be either in the elemental state or as organic matter, but not as carbonate. The investigators did not identify graphite as the form of elemental carbon by any laboratory tests but inferred its existence implicitly.

The process of formation is similar to that invoked to explain elemental carbon in meteorites. The products of the reaction are favored by decreasing temperature and increasing pressure. The hypothesis assumes that CO is an important constituent of magmatic vapors.

Metamorphic Rocks

The mineral graphite, is varying degrees of disorder is found to occur naturally. In the weathering cycle it can associate with black carbons. Mantell (1968) indicates three types of the mineral: disseminated flake, crystalline or plumbago, and amorphous or black lead. The flake variety occurs in metamorphic rocks such as marble, gneisses, and schists as a scaly or lamellar form. Plumbago is found in the form of veins or pockety accumulations along the instrusive contacts of pegmatites with limestones and schists. Flake and plumbago are characterized as being fracture or fissure fillings.

The amorphous graphites occur as minute particles distributed in weakly metamorphosed rocks, such as slates or shales, or in beds consisting almost entirely of graphite, which result from the metamorphism of coal seams. These can contain up to 80–85% of graphite, whereas the rocks carry only 25 to 60%.

The origin of the graphitic carbons in rock types can in some cases be identified by the stable carbon isotopic composition (Hahn–Weinheimer and Hirner, 1981). Besides the source materials for the graphites, the isotopic exchange reactions between carbon compounds influence the isotopic ratios. The temperatures at which isotopic equilibrium took place are also important. For example, in the cases of low grade metamorphism where the source material is organic matter, the δ C-13 values range about -20 ppm. On the other hand, graphites formed from the decarbonation of marine limestone can have positive values of the δ C-13. It is important to emphasize, according to these workers, that the carbon isotopic composition alone cannot be used to identify the source materials. Other geochemical evidence from the associated mineral assemblages and this history of the formation must be utilized. For example, epigenetic graphite is produced only from syngenetic graphite or carbonaceous detritus (Weiss et al., 1981). The mechanism may involve the conversion of carbon to CO (water–gas reaction) followed by the conversion of CO to C by the Boudouard reaction. The isotopic ratios would then be influenced by those of the source material, reactions with carbonate minerals, and upon the pressure and temperature conditions of the environment.

Hydrothermal Vent Sediments

Graphitic crystals, sometimes with antimonide overgrowths, were recovered from sediment traps close to hydrothermal vents within the 13°N East Pacific Rise. The traps placed 50 m above the seafloor were in the hydrothermal plume (Jedwab and Boulègue, 1984). The investigators propose that the graphite crystals are formed during serpentization of olivine in which the products elemental iron and hydrogen react with CO or CO_2. In association with the graphite is metallic iron. This observation is strongly suggestive that the following serptenization reactions are involved:

$$8[(Mg_{1.5}Fe_{0.5})SiO_4] + 8H_2O \rightarrow 4[Mg_3Si_2O_5(OH)_4] + Fe_3O_4 + Fe$$

$$6[(Mg_{1.5}Fe_{0.5})SiO_4] + 7H_2O \rightarrow 3[Mg_3Si_2O_5(OH)_4] + Fe_3O_4 + H_2$$

OCCURRENCE OF BLACK CARBONS

Aerosols

Both marine and terrestrial aerosols contain black carbons. The highest concentrations are found in urban airs (Wolff et al., 1982). There appears to be a decreasing gradient going from large cities to remote areas (Table 6.1). On the other hand, a gradient is not so evident for the organic carbon, suggesting that much of this material is of natural origin. For the emissions from fossil fuel burning, the black carbon/total carbon ratio is around 0.35. Thus, where the ratio has low values, as in remote and rural locations, one induces that there are large contributions of natural organics.

A general description of black carbon in atmospheric aerosols has been proposed by Clarke et al. (1984):

Location	Black Carbon (ng/m^3)
Stratosphere	0–10
Upper troposphere, maritime	4–15
Boundary layer	3–30
Continental tropospheric background	100–2000
Urban industrial areas	1000–20,000

It should be emphasized that there can be stratifications of the black carbon in a particular atmospheric zone (Hansen et al., 1984; Schnell and Raatz, 1984). For example, in the Norwegian Arctic there were three tropospheric layers on 31 March 1983: at 1 km, at 2.5 km, and at 4.5 km. The first two had concentrations equal to or greater than those at ground level (Rosen et al., 1984).

The detailed continental patterns were developed by Weiss and Wagonner (1982) who investigated 15 locations for the black carbon contents of aerosols by two methods, light absorption and single scattering albedo measurements. The black carbon fractions of the aerosols averaged 20% in industrial urban areas and 10% or less in remote areas of the United States, Hawaii, and the USSR. The reasons for the lower values of Weiss and Wagonner, compared to those of Wolff et al., are not known. The contribution of the black carbon to visibility reduction by particulates in these areas studied by Wolff et al. (1982) varied between 18 and 48%.

Atmospheric dusts are darkened through their contents of black carbons as well as those of fly ashes (Parkin et al., 1970). In some initial studies of dusts collected in the Barbados, the jet black powders that remained in suspension, following repeated ultrasonic treatment and centrifugation, were identified as carbon.

In a cruise between Ireland and Newfoundland in 1969, the presence of dark spherules in the airs was presumed to be indicative of urban pollution. Parkin et al. (1970) argued that the crumblike particles arose

Table 6.1. Black Carbon and Organic Carbon in Atmospheric Particulates (Wolff et al., 1982)[a]

Location	Black Carbon ($\mu g/m^3$)	Organic Carbon ($\mu g/m^3$)	Black Carbon / Total Particulates	Black Carbon / Total Particulate Carbon
Urban				
New York City	13.3	19.8	0.09	0.04
Washington	6.5	5.1	0.11	0.56
Denver (total)	5.4	10.4	0.06	0.34
Denver (fines)	4.4	7.6	. . .	0.35
Downey	4.1	5.9	0.06	0.41
Suburban				
Warren	3.7	8.6	0.06	0.29
Pleasanton	3.2	6.4	0.03	0.33
Pomona	3.6	8.0	0.04	0.31
Rural				
Abbeville	1.7	10.8	0.04	0.14
Luray	1.7	7.7	. . .	0.18
Remote				
Pierre				
(total)	1.1	5.1	0.08	0.18
(fines)	0.8	3.4	. . .	0.19

[a]Reproduced with permission of the authors and Plenum Press.

from diesel engines and the spherical ones (cokey balls) from oil burning. Of importance is their observation of the ubiquity of such particles in airs collected over marine waters.

The persistence or lifetime of black carbon particles in the atmosphere is controlled by four factors (Ogren and Charlson, 1983): the initial size distribution, the concentration of ambient particles, the frequency and duration of precipitation episodes, and the efficiency of removal mechanisms. Based upon models of particulate black carbon, the atmospheric residence times are calculated to vary from under 40 hr in rainy climates to well over 1 week in clean, arid regions. For particles in situations where there are high atmospheric persistences, long-range transport is possible. The high concentrations of black carbon. presumably, anthropogenic, in Arctic aerosols illustrates this point (Rosen et al., 1982). The black carbons achieved values of 40% of the total carbon content in atmospheric particulates collected at Barrow, Alaska. These levels are similar to average values for Los Angeles and New York.

In Seattle, Washington, wet removal accounted for 52–93% of the black carbon deposition and for over 99% in rural sites in Sweden (Ogren et al., 1984). In only half of the 12 Swedish sites were values for dry deposition removal higher than 15%. Using the field data at these sites, atmospheric residence times could be computed on the basis of an assumed mixing height. For heights of 1 km and 100 m, in Washington the residence times were 1 month and 3 days, respectively. A 1-km height in Sweden gives a residence time of 1 month. Median concentrations of black carbon in the rainwaters were 60 and 100 μg/L for Seattle and Sweden, respectively. Highest concentrations were associated with the lowest rainfalls.

A similar sense is provided by Muller (1984) for Middle European conditions where a residence time of about 4.5 days for black carbon is computed. Here, about two thirds of the black carbon is removed by precipitation.

Additional support for long time persistence can be seen in the black carbon contents of air masses collected on the Louisiana Gulf Coast (Wolff et al., 1983). Two types of airs were studied: those that had background levels in which the air had resided over the Gulf for several days—the so-called maritime tropical air masses; and those in which the air mass had traveled through the midwestern and northeastern United States to the southeast and then to the southwest over the Gulf of Mexico. The latter airs had higher average black carbon concentrations (2.6 $\mu g/m^3$), attributable to the long-range transport of emissions from the midwestern and northeastern United States, than the former airs (black carbon concentrations of 1.2 $\mu g/m^3$).

Black carbon in urban aerosols is usually found in the fine size fractions. It constitutes about 50% or more of the noncarbonate carbon (Budiansky, 1980 and Novakov, 1982). Sulfur when found in the aerosols is associated with the black carbon clusters (Russell, 1979). Some values for U.S. cities are given in Table 6.2.

Table 6.2. Black Carbon Concentration in Aerosols of Some U.S. Cities (Novakov, 1982)[a]

City	Black Carbon (%)
New York	97
Gaithersburg	81
Argonne	74
Berkeley	73
Anaheim	68
Fremont	64
Denver	59

[a] Reproduced with permission of the author and the American Geophysical Union.

Concentrations were usually above 1 $\mu g/m^3$ at both urban and rural sites. Black carbon concentrations ranged from 1 $\mu g/m^3$ in a remote location in South Dakota to 13.3 $\mu g/m^3$ in New York, in a study of airs sampled at 10 sites in the United States. About 80% of the black carbon is found in the fine particles, $<2.5\ \mu m$ in diameter. The total carbon:black carbon:lead ratio in ambient airs usually runs 9.4:3:1 for the Los Angeles air basin. Total carbon to elemental carbon usually is constant at about 3:1. The ratio, total carbon to elemental carbon, varies from 1.3:1 for diesel cars to 2.3:1 for catalyst equipped nondiesels.

In Hamburg, Federal Republic of Germany, the average black carbon concentrations in the atmosphere during a 17-month period in 1981–1982 were 2.4 $\mu m/m^3$. This represented about 4% of the total suspended particle content. During winter smog episodes, the black carbon content attained values of 24 $\mu g/m^3$ (Heintzenberg and Winkler, 1984). The investigators attributed sources for the carbon to diesel cars and stationary combustion sources with the former apparently dominating.

Black carbons in aerosols collected in the Great Smoky Mountains of Tennessee at an elevation of 646 m were studied by Stevens et al. (1980). Two size classes of aerosols were obtained by filtration techniques: 0–2.4 or 2.5 μm and 2.4 or 2.5 to 15 or 20 μm. On the average the fine fractions represented about 80% of the aerosol mass. Black carbon represented

5% of the total mass of the fine particles, whereas, organic carbon accounted for 10%. The black carbon concentration in the fines was 1100 ng/m^3. Sulfate and its cations represented 61% of the fine particle mass, approximating the chemical form of ammonium bisulfate. The study carried out in 1978 indicated that the fine-particle aerosol was dominated not by natural organic compounds but by acid sulfates and charcoal, associated with fossil fuel combustion.

Ogren (1982) analyzed aerosol samples that had deposited on the University of Washington campus between 25 April and 12 May 1980. The samples were collected in 1-L glass beakers, containing about 40 mg of mercuric chloride acting as a biocide and subsequently passed through a 5-μm nylon mesh onto 25-mm glass fiber filters. An automatic sampler exposed only one sample at a time, depending upon whether or not it was raining. The amounts of charcoal were determined by optical absorption measurements, not specific for black carbon. Ogren (1982) indicated that the black carbons are the dominant species in urban aerosols on the basis of the optical studies of Rosen et al. (1978) and that his results are indicative of true black carbon concentrations.

The average black carbon concentration in the air during the sampling period was 0.5 μg/m^3 with measured depositional fluxes of 9.0, 7.0, and 0.36 μg/m^2/hr for total, wet and dry removal, respectively. The corresponding residence times then are 3.5, 4.5, and 87 days for total, wet and dry deposition. Similar values have been found for sulfate aerosols and the removal mechanisms for charcoal and the sulfates may be similar. The charcoal concentration in the rains was 240 μg/L, which yields a washout ratio of 5×10^5. The dry deposition velocity is computed to be 0.02 cm/s. On the basis of these results clearly wet deposition is the important removal mechanism.

An investigation into the nature of the haze in the Denver, Colorado, USA area indicated that black carbon was an important component of the particulate matter (Heisler et al., 1980). The total suspended atmospheric solids were divided into a fine fraction (particle diameters <2.5 μm). The major constituents of the fine particles were organic materials (17–35%); sulfate (8–19%); black carbons (9–17%); nitrate (6–14%), and crustal debris (5–17%) (see Table 6.3). The carbon was the prime contributor to the particle light extinction coefficient for the haze through absorption and scattering and accounted for 38% of it. Only the fine fraction influenced the nature of the haze. The coarse particles, about equal in weight to the fine particles, contribute primarily to the "total suspended particulate" measurement. This fraction is composed primarily of crustal debris.

Somewhat different results were obtained by Groblicki et al. (1981) for the Denver haze. Rayleigh scattering accounted for 7% of the total

Table 6.3. Composition of the Fine and Coarse Particulates in the Denver Haze (%) (Heisler et al., 1980)[a]

Component	Fine	Coarse
SO_4^{2-}	8–19	0.3–4
NO_3^-	6–14	1–3
NH_4^+	5–8	0–2
Black carbon	9–20	1–4
Organic substances	17–35	4–13
Nonsulfate S	0–2	0.4–2
Pb + Br	3–6	2
Cl	1–2	2–3
Trace metals[b]	0.2–0.4	0.2–0.3
Water[c]	5	5
Crustal debris	5–17	64–86

[a]Reproduced with permission of the authors from "The 1978 Winter Haze Study," prepared for the Motor Vehicle Manufacturers Association of the United States by Environmental Research and Technology, Inc.
[b]Usually included, V, Cr, Mn, Ni, Cu, Zn, As, and Se.
[c]Assumed to be 4.8% of mass.

extinction. Contributions to visibility reduction were from ammonium nitrate (17%), organic compounds (13%), elemental carbon (38%), ammonium sulfate (20%), other particulate matter (7%), and NO_2 (6%). Most of the visibility reduction was attributed to particles smaller than 2.5 μm, with minor contributions from elemental carbon particles larger than 2.5 μm.

Atmospheric particulates collected during the summer of 1977 in St. Louis were found to have an average concentration of 65 $\mu g/m^3$ of which 12% was carbon, as determined by reflectance measurements (Delumyea et al., 1980). Black carbon constituted about half of the carbon content.

By the same techniques, black carbons were measured under wintertime conditions in the Los Angeles area (Pasadena, California and downtown Los Angeles, California) (Conklin et al., 1981). The black carbon averaged about one third of the total carbon and declined in concentration from early morning until midday corresponding to the changes in hourly traffic density.

Using the charcoal concentrations in Los Angeles air (9 $\mu g/m^3$) and the absorption efficiencies, the authors calculated that 7 to 17% of the total light extinction could be attributed to the black carbon.

The role of black carbon on atmospheric visibility impairment in other larger cities has also been estimated. Chu and Macias (1983) indicate that the relative contributions to visibility impairment in St. Louis, Missouri are 49% by the scattering of sulfates, 14% by the absorption by black carbon, 6% by the scattering by black carbon, 11% by the scattering of organic carbon, and 12% by the scattering of the remainder of the ambient fine aerosols. Black carbon accounts for 6% of the fine mass but for 20% of the visibility reduction.

In Vienna, Austria, the absorption coefficient of the aerosols varies between 10% in summer to 50% in winter (Habenreich and Horvath, 1984). There is a covariance between traffic density and light absorption indicating that traffic is an important source of the black carbon.

The seasonal variation of black carbon in Arctic aerosols has been studied by Rosen et al. (1981) who found there was a marked increase in content from late fall to early spring during 1979–1980. Further, the black carbon percentage in the total carbonaceous particulates rose during this time period and reached values of nearly 40%, a value higher than that found in such urban centers as New York City and Los Angeles. These patterns were also observed at many locations in the Arctic and subarctic regions. Since forest fires are not extensive during this time period, the black carbons found in the aerosols are attributed to long-range transport of fossil fuel combustion products from industrialized areas.

The allocation of the contributions of various particle types to the extinction coefficient in Denver haze was made first by relating the extinction coefficient to the concentrations of the various types of particles from different sources and, secondly by ascertaining the contribution of emissions to atmospheric concentrations. The composition of the fine particles, which dominate particle scattering, and the course particles, which are involved with absorption, are given in Table 6.3. The compositions of fine particles with respect to elemental carbon and organic carbon, as well as those of other constituents (Table 6.4) were obtained both from the literature and from new analyses made in the laboratory of the investigators (Heisler et al., 1980). The allocations are given in Table 6.5.

An extensive study of atmospheric particulates in the Denver, Colorado area between November 31 through December 23, 1978

Table 6.4. Composition of Fine Particles (%) (Heisler et al., 1980)[a]

Material	Elemental Carbon	Organic Carbon
Auto exhaust, unleaded	43.3 ± 17.3	36.1 ± 14.4
Auto exhaust, leaded	7.0 ± 4.3	21.6 ± 15.0
Diesel exhaust	54.0 ± 33.0	38.0 ± 22.0
Residual oil	3.1 ± 2.5	7.0 ± 6.2
Coal[b]	0	0
Natural gas	51.0 ± 21.0	41.0 ± 16.0
Soil dust	0.6 ± 0.4	4.3 ± 1.7
Crustal material	0.3 ± 0.07	2.2 ± 0.5

[a]Reproduced with permission of the authors from "The 1978 Winter Haze Study," prepared for the Motor Vehicle Manufacturers of the United States by Environmental Research and Technology, Inc.
[b]No explanation is given for this unusual result.

Table 6.5. The Allocation of the Particle Extinction Coefficient for the Denver Haze to Various Emission Sources (Heisler et al., 1980)[a]

Source	Percent of Extinction Coefficient
Automobiles–catalyst	3.6
Automobiles–noncatalyst	10.2
Diesel vehicles	15.4
Natural gas combustion	13.4
Crustallike material and coal combustion	13.4
Residual oil combustion	2.9
Unidentified charcoal sources	15.0
Unidentified organic carbon sources	10.4
Water	3.5
Unallocated	12.2

[a]Reproduced with permission of the authors from "The 1978 Winter Haze Study," prepared for the Motor Vehicle Manufacturers of the United States by Environmental Research and Technology, Inc.

revealed an apparent black carbon concentration in the airs (based upon the General Motors analytical technique, see Appendix) ranging from 0.2 to 35.7 $\mu g/m^3$ (Wolff et al., 1980). The average was 6.6 $\mu g/m^3$.

The black carbon concentrations displayed a strong correlation with the particulate organic carbon (correlation coefficient of 0.97) and the next strongest correlations were with the vehicular exhaust components (Pb, Cl, and N) ($r = 0.89 - 0.91$). This latter result is in accord with the diurnal fluctuations of the black carbon and of the total particulate carbon, which show peak concentrations associated with the 0800 to 1200-hr period and minimum concentrations in the 1200–1600-hr period. These maxima and minima conform to vehicular traffic patterns.

There are also strong correlations with the coal and oil combustion emission products such as As, Si, and V ($r = 0.89$–0.91) in the fine fractions of the aerosols. The other sources of elemental carbon do not have unique particulate tracers and hence any relationships with them (natural gas combustion, wood burning) cannot be made.

Much of the black carbon in urban atmospheres can originate from wood burning. Griffin and Goldberg (1979) examined the larger particles in air from La Jolla, California, and gave the following composition for them, based upon their morphologies: 36% wood carbon, 16% oil carbon, 5% unidentifiable carbon, and 45% pollen grains. A similar result comes from C-14 analyses of atmospheric carbonaceous parti-

Table 6.6. Estimated Source Contributions of Fine Black Carbon in the Denver Area (Wolff et al., 1980)[a]

Source	Daily Emission Rate (kg/day)	Percent of Total Charcoal
Light-duty catalyst vehicles	47	1.5
Light-duty noncatalyst vehicles	164	5.3
Diesel vehicles	632	20.4
Natural gas	514	16.6
Fuel oil	217	7.0
Coal	19	0.5
Aircraft	199	6.4
Wood	1195	38.5
Total of known sources		96.2
Unknown sources		3.8
Total		100

[a]Reproduced with permission of the authors and Plenum Press.

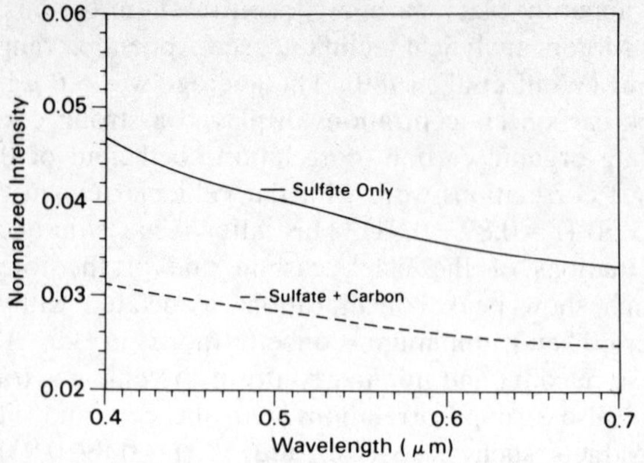

Figure 6.3. Horizon sky intensity as a function of wavelength (Bergstrom et al., 1982). See text for explanation of curves. Reproduced with permission of the authors and Plenum Press.

culates from Denver, Colorado (Wolff et al., 1980). Carbon with a modern C-14 activity can derive from the burning of wood, vegetation, and refuse. The samples were collected in November and December when vegetation burning is minimal. Further, refuse incineration is prohibited. Thus, the high C-14 activity in the samples is attributed to wood burning. For the five samples analyzed, 30–43% of the particulate carbon, including the black carbon had contemporary C-14 activities. Muhlbaier and Williams (1982) had previously shown that about 80% of the carbon particulates from wood burning fireplaces are <2.5 μm in size and that the charcoal carbon to organic carbon had a ratio, 49/51. Using these numbers Wolff et al. (1980) estimate that 39% of the black carbon in the atmosphere had an origin in wood burning, a number quite similar to that found in La Jolla (Table 6.6).

The impact of sulfate and black carbon aerosols upon the horizon sky intensity of an urban area has been computed by Bergstrom et al. (1982). Two models are formulated: one assumes that all of the aerosol in the nonabsorbing case is "sulfate only" and the second assumes that there is 5 $\mu m^3/cm^3$ of black carbon with a geometric mean radius of 0.01 μm. The carbon substantially reduces the sky intensity at all wavelengths, with the blue being somewhat more effective than the red (Figure 6.3).

Ice Nuclei

Black carbon particles were found associated with ice nuclei collected from the airs of College, Alaska and Nagoya, Japan (Isono et al., 1971).

Air was introduced into a cloud chamber that was cooled down to −20°C and nuclei grew to sizes of about 50 μm in diameter. The nuclei were then collected upon a filter paper and the water allowed to evaporate.

At these locations, the ice nuclei were mainly clay minerals. Around them, however, black carbon particles were found. The carbon particles did not appear to be the nucleating agents. Their sizes were usually under a micron and they in some cases had spherical shapes.

There were no evident black carbon particles in snow crystals sampled at the South Pole by Kumai (1976). This observation may result from a lack of extensive industrial activity and biomass burning in the higher latitudes of the southern hemisphere.

Sediments

Marine, lacustrine, and probably glacial sediments maintain histories of biosphere (plant and wood) burning, both natural and anthropogenic. This was first pointed out by Smith et al. (1973) through an ivestigation of the black carbon contents of open ocean Pacific sediments. The deposits were accumulating at rates of the order of 1 mm/1000 yr. Samples were taken from the upper 15 cm and hence the results reflect averages of black carbon concentrations over the past 150,000 yr or so.

There is a marked increase in the black carbon concentrations of the pelagic sediments going from the southern equatorial regions northward (Figure 6.4). This latitudinal variation corresponds to the zonation of nontropical forest types about the earth's surface. The types of forests in this domain depend upon soil type, climate, elevation, and past land use. At high latitudes before the ice sheets are approached, the tundra is covered by mosses and lichens with only occasional small trees. Cool coniferous forests belt the Scandinavian countries across Russia to Siberia and continue over much of northern Canada. Their trees consist largely of spruce, pine, fir, hemlock, and cedar with some birch and willow. Southward, they blend into the temperature mixed forests of the eastern United States, the British Isles, central Europe, and eastern Asia. Fire is initiated as a consequence of lightning strikes. The charcoal is produced as a result of the incomplete combustion of the plant tissues.

The tropical forests span the near-equatorial regions. In the Amazon basin, western and west central Africa, and in southeastern Asia there are high concentrations of these forests. In general, the nonanthropogenic combustion of tropical forest organic matter is through low temperature, biologically mediated processes without the production of charcoal. In contrast to these rain forests are the drier open woodlands or savannas,

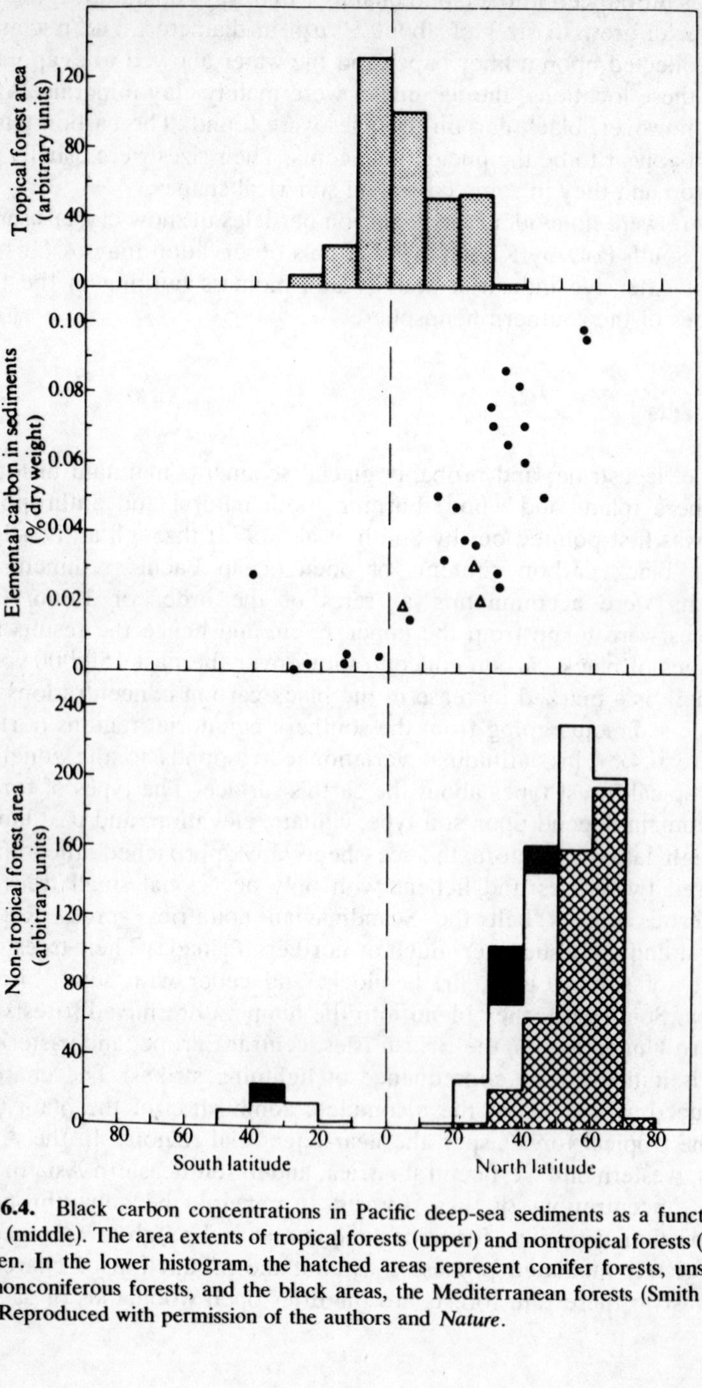

Figure 6.4. Black carbon concentrations in Pacific deep-sea sediments as a function of latitude (middle). The area extents of tropical forests (upper) and nontropical forests (lower) are given. In the lower histogram, the hatched areas represent conifer forests, unshaded areas, nonconiferous forests, and the black areas, the Mediterranean forests (Smith et al., 1973). Reproduced with permission of the authors and *Nature*.

which contain primarily grasses and only small numbers of trees. This part of the biosphere can also burn at high temperatures producing charcoals.

The winds that can move the debris of high temperature burning from the continents to the marine environment are zonal. The three chief systems in the lower atmosphere are the easterly trades of the equatorial regions which blow from east to west, the westerlies of the midlatitudes, much more intense at higher elevations and more variable than the trades, and the polar easterlies of relatively low intensity and high variability.

Rain

The association of black carbon with other industrial pollutants, such as hydrogen ion and sulfate ion was first pointed out by Gorham (1955) in the study of 42 rain samples collected in the English Lake District in

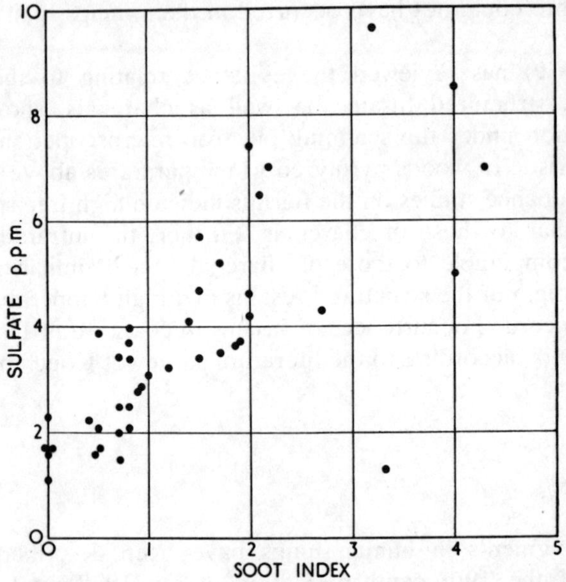

Figure 6.5. The relationship between the sulfate concentrations and black carbon (measured through darkening of filter papers and called the soot index) in English Lake District (Gorham, 1955). Reproduced with permission of the author and Pergamon Press, Ltd.

1954. Gorham (1955) filtered his rain samples and measured the blackness of the filters. He graded them on a scale of 0 to 5 and obtained a measure of "soot" by dividing the grade by the sample volume in liters.

Gorham (1955) indicated that sulfuric acid rains had high black carbon concentrations (Figure 6.5). All but one of the 10 rains with the highest amount of soot had high sulfate concentrations (>4 ppm). There appeared to be a seasonal dependence upon black carbon and sulfate concentrations in rains, with decreases occurring from May to October. Hydrogen ion concentrations varied little with a maximum in summer.

Coals

Besides the sedimentary records, histories of natural combustion processes may be found in coals which contain charcoallike material, fusain, similar in many ways to those of naturally occurring carbons. There has been a dispute in the literature as to whether the coal particles are formed by high temperature processes or not. On one side, investigators base their arguments on the resemblance of the coal charcoals to those formed by the burning of biomass. On the other hand, other investigators submit that fire could not have occurred in the swamps that gave rise to the coals.

Cope (1979) has reviewed the evidence relating to the origin of fusains. The structured fusains, as well as charcoals, show cell wall homogenization under the scanning electron microscope; such features are characteristic of woods pyrolyzed at temperatures above 300°. Electron spin resonance studies on the fusains indicate high free spin concentrations, similar to those of charcoals. Further, the infrared spectra of fusains are comparable to those of charcoals. Such similarities strongly support an origin of the structured fusains in a high temperature process. The oldest recorded occurrences of fusains in coals are of Devonian age in Pennsylvania, according to the literature survey of Cope and Chaloner (1980).

Anthrosphere

The black pigments in human lungs have been described since the beginning of the 19th century (Slatkin et al., 1978) and have been identified as black carbons. Ancient human remains show evidences of black carbon deposits in the lungs. For example, the uncircumcized Egyptian mummy Pum II had charcoals in the fibrotic areas and such a finding is in accord with other examples where mummified bodies show a record of air pollution associated with burning (Cockburn et al., 1975). Most probably the charcoals arose from managed fires in confined spaces such as caves or huts.

The black deposits have a C-13/C-12 ratio distinctly lower than that in the endogenous lung tissues (Slatkin et al., 1978). For a peripheral lung parenchyma extracted from 20 New York autopsies the mean of the $\delta^{13}C$ value from the peripheral lung parenchyma was -16.0 ± 0.7 while that from the black pigments (carbon) in the carinal or bronchial lymph nodes was -22.2 ± 2.7. The black carbon represents about 30% by weight of the pigmented lung material.

7

CHEMICAL REACTIONS INVOLVING BLACK CARBONS

INTRODUCTION

Black carbons in the environment may be involved in chemical reactions primarily as a consequence of the catalytic activity of their surfaces. Although some work has been done along this line with atmospheric gases, the involvement of black carbons with reactions in the sedimentary column has not been investigated but remains a strong possibility.

The surface functional groups and the nature of the carbon itself govern the types of electron transfer and hydrolysis reactions that can occur. Surface defects and discontinuities probably play important roles in catalyzing reactions.

The surface functional group can themselves be governed by exposure to atmospheric gases and perhaps be involved in chemical reactions. Akhter et al. (1984c) have shown that hexane soots exposed to NO_2 at partial pressures between 64 and 240 torr produced novel functionalities including $-C-NO_2$, $-C-ONO$, and $C-N-NO_2$. The involvements, if any, in the chemistries to be discussed are yet to be discovered.

REACTIONS

Nitrogen- and Sulfur-Containing Gases

Heterogeneous reactions involving black carbons with atmospheric gases have been extensively studied over the past decade. Much attention has been paid to the oxidation of sulfur dioxide on the surfaces of the carbon particles with ozone, nitrogen oxides, and light. Black carbons appear effective in catalyzing such reactions because of the nature of their surfaces and because of their abundances.

Ammonia reacts very rapidly (a probability of a reaction of >0.001/collision) with soot formed by condensation from a methane diffusion flame. The reaction rate decreases substantially with exposure of the particle surface. Some species, acting alone, showed no reactivity (a probability of reaction < 1/million collisions). Among them are NO, NO_2, SO_2, HNO_3, H_2O_2, and N_2O_5.

The occurrences of reduced species of nitrogen, amines and/or amides, and nitrile, were first noted in atmospheric aerosols by Novakov et al. (1972). The variations in their concentrations were diurnal and indicated an origin in pollutant sources such as automobiles rather than in the atmosphere. Their presence in atmospheric aerosols is ubiquitous. Subsequently, Chang and Novakov (1975) have shown that these compounds can be produced by the interactions of black carbon with NO or NH_3. The work was carried out using the ESCA (electron spectroscopy for chemical analysis) technique. An ESCA spectrum of ambient atmospheric particles is shown in Figure 7.1. The individual peaks for NO_3, NH_4, and N_x (the reduced nitrogen species) are evident. Most of the nitrogen is found as

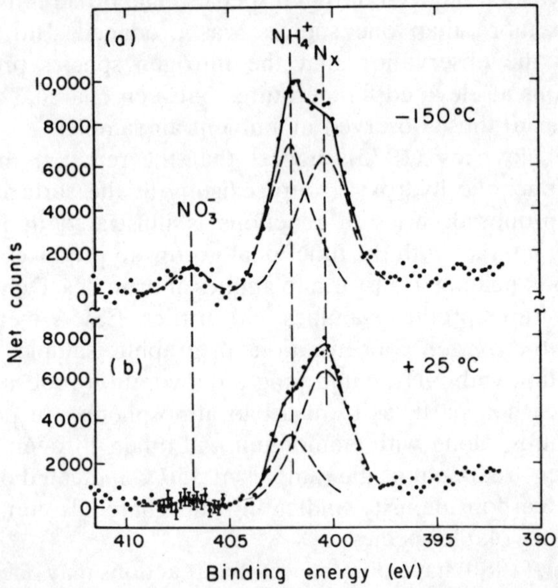

Figure 7.1. (*a*) Nitrogen (1*s*) photoelectric spectrum of atmospheric particulates at −150°C. NO_3^-, NH_4^+ and N_x peaks are indicated. (*b*) The spectrum at 25°C (Chang and Novakov, 1975). Reproduced with permission of the authors and *Atmospheric Environment*.

the ammonium ion. The peak areas are proportional to the relative atomic concentrations, as nitrogen, of these species. The samples were initially analyzed at −150°C to minimize the losses of volatile species. When the sample temperature was raised to 25°C, there was a significant decrease in ammonia (60%) (Figure 7.1). The amount of ammonia lost during this temperature change was far greater than the amount of original nitrate present indicating that NH_4NO_3 is not the volatile species. The ammonium salt was not identified although ammonium sulfate can be eliminated inasmuch as it is stable to volatilization over the time period in which the analyses were carried out.

Chang and Novakov (1975) carried out laboratory experiments to ascertain the products from the interactions with air of freshly prepared soots from a propane/oxygen flame. Two types of reactions were studied: (1) a static one in which the airs interacted with the charcoal deposited on a silver membrane filter at ambient temperatures and (2) a flow system in which the reactant gas was introduced downstream from the propane/oxygen flame, allowing the soot particles to be at high temperatures. Under both conditions the charcoal was exposed to NO and NH_3. At low temperatures, the interactions of charcoal with NO and NH_3 produced ammonium-containing species. On the other hand, at high temperatures, the reactions yielded reduced nitrogen species. The broad nitrogen peaks indicated that more than one species was produced. But of greater importance is the observation that the nitrogen species produced by surface reactions at elevated temperatures between charcoal and NO or NH_3 are similar to those observed in ambient air samples.

Chang and Novakov (1975) suggest that the reactions involve the carboxyl and phenolic hydroxy groups existing at the surfaces of black carbon. The proposed series of reactions is illustrated in Figure 7.2. Ammonia may interact with the functional groups to produce ammonium salts which, upon heating, transform to amines and amides. Dehydration at higher temperatures produces amines and nitriles. ESCA measurements indicated that the oxygen concentrations of graphite samples are higher before interaction with NH_3, confirming a dehydration process.

The black carbon particles from urban atmospheres are produced at high temperatures, along with ammonium and other nitrogen containing species. Further, treatment of the samples at 350°C indicated only a small loss in ESCA carbon intensity, indicating the charcoals contain a very small amount of volatile species.

Schryer et al. (1980) have pointed out that reactions may occur through synergistic effects among atmospheric trace gases. They noted no chemisorption of either SO_2 or NO_2 in dry air or nitrogen gas upon carbon black. However, when both gases are exposed together to the carbon

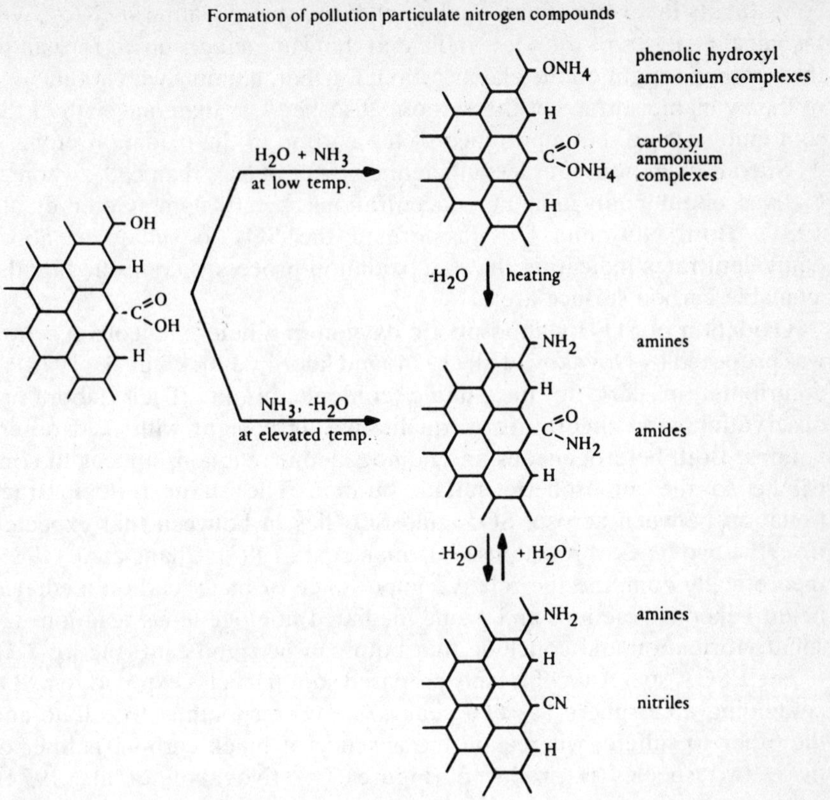

Figure 7.2. Postulated reactions occurring at the surfaces of black carbon particles (Chang and Novakov, 1975). Reproduced with permission of the authors and *Atmospheric Environment.*

black, a significant chemisorption results with the sulfur dioxide being converted to sulfate. The NO_2 seemed to be the oxidizing agent as the reaction took place both in air and nitrogen gas. Some oxidation of SO_2 to sulfate occurs in humid air without the presence of NO_2; NO_2 does increase the yield by an order of magnitude or more.

There appeared to be no quantitative chemisorption on carbon black under the following conditions: SO_2 in dry air, in dry N_2, or in humidified N_2 in the absence of NO_2; or with NO_2 in these carriers in the absence of SO_2.

The effectiveness of NO_2 is converting SO_2 to sulfate in the presence of carbon black at relative humidities of 65% has been demonstrated by Cofer et al. (1980). Sorption of the gases was stronger in nitrogen atmospheres than in air suggesting that molecular oxygen or some trace constituents in air may have weakly inhibited the oxidation by NO_2. Wet chemical analyses of the soots indicated that sulfate accounted for half of the retained weight on the black carbon. Further, alumina with about 54% of the available surface of the carbon, absorbed less than one sixth of the soot emphasizing the importance of the carbon in the oxidation process.

Nitrous oxide was a far less efficient oxidizer of SO_2 than NO_2, whereas O_3 was about equivalent at concentrations ≥ 0.07 ppm (Cofer et al., 1981). Both NO_2 and O_3 transformed the SO_2 to sulfate at about equivalent rates indicating that the oxidation process is controlled by the available carbon surface areas.

Oxidation of SO_2 by atmospheric oxygen in a heterogeneous reaction was proposed by Novakov et al. (1974) and later by Judekis et al. (1978) to contribute, in part, to the sulfate contents of air. Their laboratory observations and theory are in qualitative agreement with field observations. Both heterogeneous and homogeneous reactions appear to contribute to the atmospheric sulfate burden. The sulfur isotopic fractionation between aerosol SO_4^{2-} and SO_2 lies in between that expected from the two types of reactions (Saltzman et al., 1983). Chang et al. (1981) theoretically compute the relative importance of black carbon mediated heterogeneous reactions and ozone mediated homogeneous reactions for sulfate formation and conclude that both can be significant (Figure 7.3).

The ESCA spectrum of freshly prepared soot particles exposed to a SO_2 containing atmosphere has two peaks, one corresponding to sulfate and the other to sulfate, whereas in the absence of black carbon, neither of these two species is produced (Figure 7.4) (Novakov et al., 1974). Subsequent experiments involved the exposure of dried and humidified airs to soots. The air containing water always produced more intense peaks, whereas when N_2 was substituted for air under both dry and humid conditions, only background levels of sulfate were found. This indicates that the oxidation involves molecular oxygen and not water (Figure 7.4).

In this study the amounts of sulfate produced when the sulfate dioxide was introduced downstream of the flame producing the soot, in comparison with that produced in the flame envelope were similar, suggesting that there are only trivial, if any, contributions of sulfate from homogeneous gas phase oxidation. The soot was produced in a C_3H_8/O_2 flame.

The role of the water in the oxidation process appears to be that of a "getter" for atmospheric gases (Matteson, 1979). Droplets which grew by water vapor condensation in laboratory experiments concentrate such gases as SO_2, NO_2, and O_2 to levels of supersaturation. Thus, the oxidation reactions may proceed at rates greater than those based upon steady state concentrations.

The amount of SO_2 converted to sulfate is independent of the initial

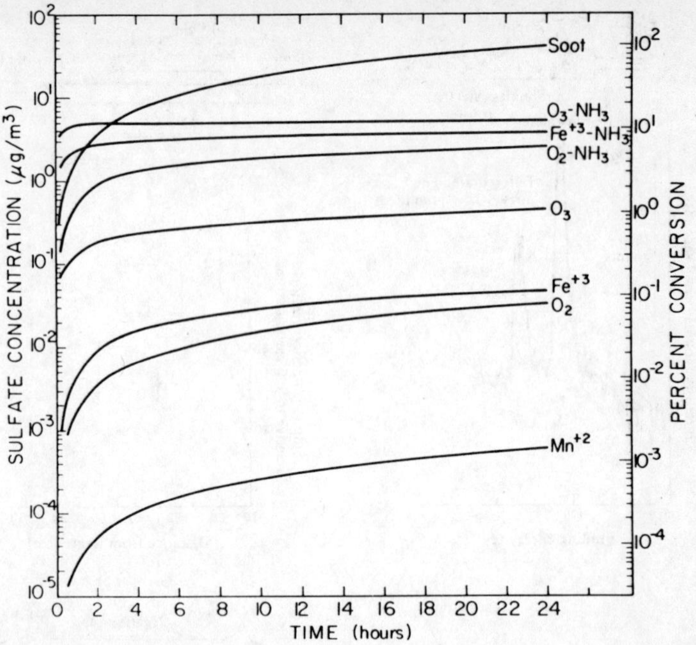

Figure 7.3. A comparison of the relative importance of various sulfate production mechanisms on the bases of model calculations. The following initial conditions were used: $P_{SO_2} = 0.01$ ppm; $P_{CO_2} = 0.000311$ atm; $P_{NH_3} = 5$ ppb; $P_{O_3} = 0.05$ ppm; $P_{HNO_2} = 8$ ppb; [Fe(III)] = 0.12 μM; [Mn(II)] = 0.018 μM; charcoal – 10 $\mu g/m^3$ and liquid water – 0.05 g/m^3 (Cheng et al., 1981). Reproduced with permission of the authors and *Atmospheric Environment*.

concentration of the reactant (Figure 7.4). Further the decrease in SO_2 activity, ΔSO_2, as a function of the O_2/C_3H_8 ratio is essentially identical for SO_2 concentrations of 5.5 and 9.9 ppm. Novakov et al. (1974) attribute this to a saturation of active sites on the black carbon that are capable of mediating the oxidation. The covariance of ΔSO_2 with the O_2/C_3H_8 ratio reflects the increase in production of soot particles produced in the O_2 rich flame.

The initial rate of adsorption of SO_2 on a fresh black carbon surface was found to be first order in SO_2 and carbon up until saturation was reached (Baldwin, 1982). The initial rate corresponds to an atmospheric loss of 1.2%/hr for a typical atmospheric burden of charcoal of 100 $\mu m/m^3$. This is comparable to the rate of the homogeneous oxidation of SO_2. However, since saturation does set in rapidly, this cannot be a major mechanism for the production of sulfate unless there is a regeneration of clean black

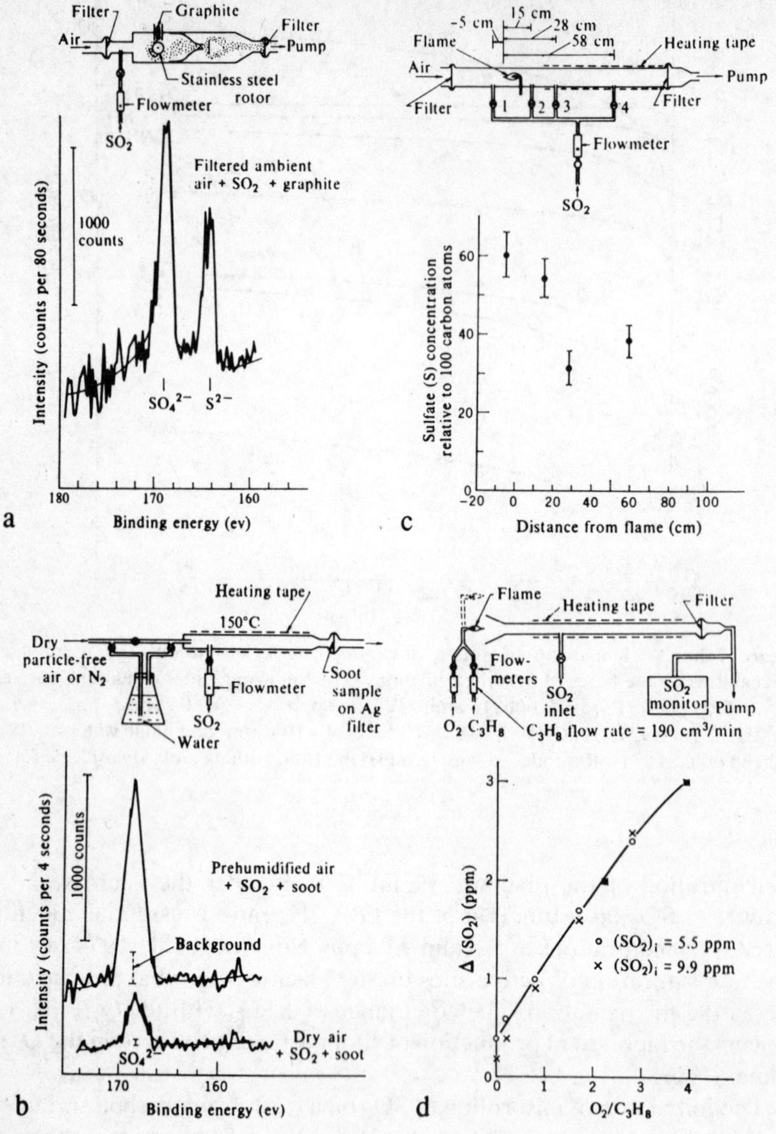

Figure 7.4. Upper: Apparatus for the exposure of soot particles to SO_2. (*a*) ESCA spectrum of charcoal surfaces revealing sulfate and sulfide peaks (see text). (*b*) ESCA spectrum of charcoal surfaces exposed to prehumidified air. (*c*) Sulfate produced at distances downflow of flame (sampling ports are in apparatus above graph). (*d*) The relationship between initial SO_2 values and the uptake of SO_2 as a function of the O_2/C_3H_8 ratio (see text). (Novakov et al., 1974). Reproduced with permission of the authors and the American Association for the Advancement of Science.

carbon surfaces. A process for regeneration may be found in the experiments of Rogowski et al. (1982) who exposed charcoals in aqueous solution to gas mixtures of SO_2, air, and NO_2. Carbon black catalyzed the oxidation of SO_2 to sulfate either by the oxygen in air or by NO_2 alone. The yields are pH independent to values as low as 1.5. But of greater importance is the observation that no saturation occurs as long as sufficient water is present. Thus charcoal particles in air with sufficient moisture may play an effective role in the heterogeneous oxidation of sulfur dioxide by NO_2 or oxygen.

The acceleration of the SO_3 formation by NO_2 goes by

$$SO_2 + NO_2 = SO_3 + NO$$

but the amount of SO_3 produced can be depressed by high NO_2 concentrations (Britton and Clarke, 1980). This has been interpreted as resulting from active site poisoning by surface nitrate formation.

For the reaction of the oxidation of sulfate dioxide to sulfate by oxygen on a soot surface, $2S(IV) + O_2 = 2S(VI)$, the kinetics have been studied by Chang et al. (1979 and 1981) with the following empirical rate equation:

$$\frac{d[S(VI)]}{dt} = k[C][O_2]^{0.69} \frac{\alpha[S(IV)]^2}{1 + \beta[S(IV)] + \alpha[S(IV)]^2}$$

where C is the carbon content, assuming it to be proportional to the effective surface area; S(IV) is the dissolved $SO_2 = HSO_3^- + SO_3^{2-}$ and the $S(VI) = HSO_4^- + SO_4^{2-}$. The rate constants are $k = 1.69 \times 10^{-5}\ mol^{0.31} \cdot L^{0.69}/g \cdot s$, $\alpha = 1.50 \times 10^{12}\ L^2/mol^2$, $\beta = 3.06 \times 10^6\ L/mol$, [C] = g carbon/L, $[O_2]$ = mol dissolved oxygen/L, [S(IV)] = total mol S(IV)/L, and [S(VI)] = total mol S(VI)/L. It is interesting to note that the rate is first order with respect to the available sites on the carbon.

A series of adsorption steps has been proposed by Novakov (1982) to explain the first-order kinetics for oxygen:

$$C + O_2 \underset{k_{-1}}{\overset{k_1}{\rightleftharpoons}} C \cdot O_2$$

$$C \cdot O_2 + S(IV) \underset{k_{-2}}{\overset{k_2}{\rightleftharpoons}} C \cdot O_2 \cdot S(IV)$$

$$C \cdot O_2 \cdot S(IV) + S(IV) \underset{k_{-3}}{\overset{k_3}{\rightleftharpoons}} C \cdot O_2 \cdot 2S(IV)$$

$$C \cdot O_2 \cdot 2S(IV) \xrightarrow{k_4} C + 2S(VI)$$

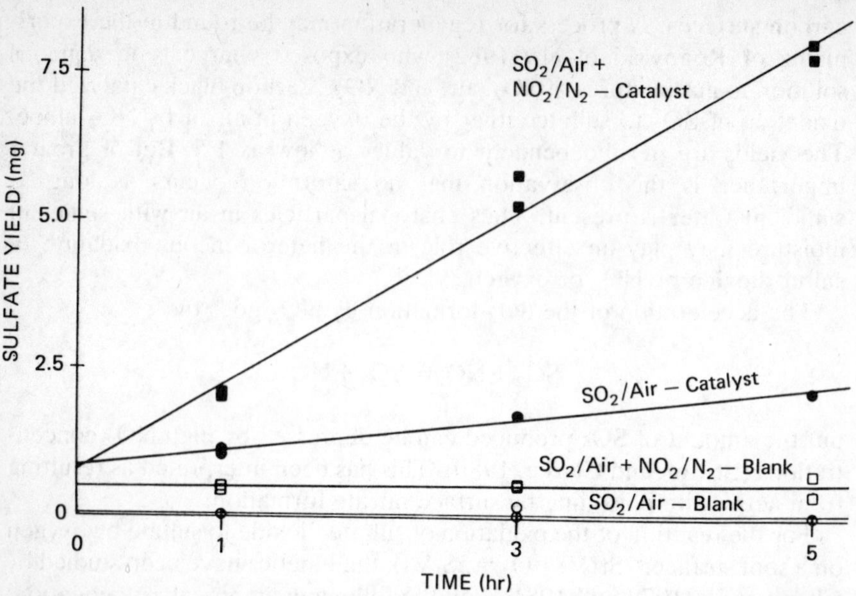

Figure 7.5. Production of sulfate from SO_2 under various conditions. Note importance of black carbon catalyst (Regowski et al., 1982). Reproduced with permission of the authors and the American Geophysical Union.

Cloud chamber experiments using dispersed droplets that contain carbon black emphasized that the solid phase was essential for rapid conversion of sulfur dioxide to sulfate (Figure 7.5) (Benner et al., 1982). The production of sulfate for the lowest concentrations of SO_2 exposed to carbon black (about 0.007 ppm) produced more sulfate than did water droplets exposed to 222 ppm SO_2.

Still there are several lines of evidence that relate the laboratory formed sulfates to those observed in the field. The sulfates arising from fossil fuel combustion are believed to be H_2SO_4 and/or $(NH_4)_2SO_4$. The sulfate produced in the laboratory is water soluble and could conceivably have been neutralized by ammonia.

The relative importance of the heterogeneous and homogeneous reactions for sulfate production have been determined by Middleton et al. (1982) on the bases of models utilizing the rates of photochemical reactions, vapor condensation, and catalytic and noncatalytic oxidations on wetted aerosols. Their results indicate that the black carbon catalyzed oxidation makes a significant contribution to sulfate production (Figure 7.6a and 7.6b) during summer and winter days and summer night. In neither daytime example is the black carbon catalyzed heterogeneous reaction dominant, however, along with other heterogeneous reactions it may be dominant during the night.

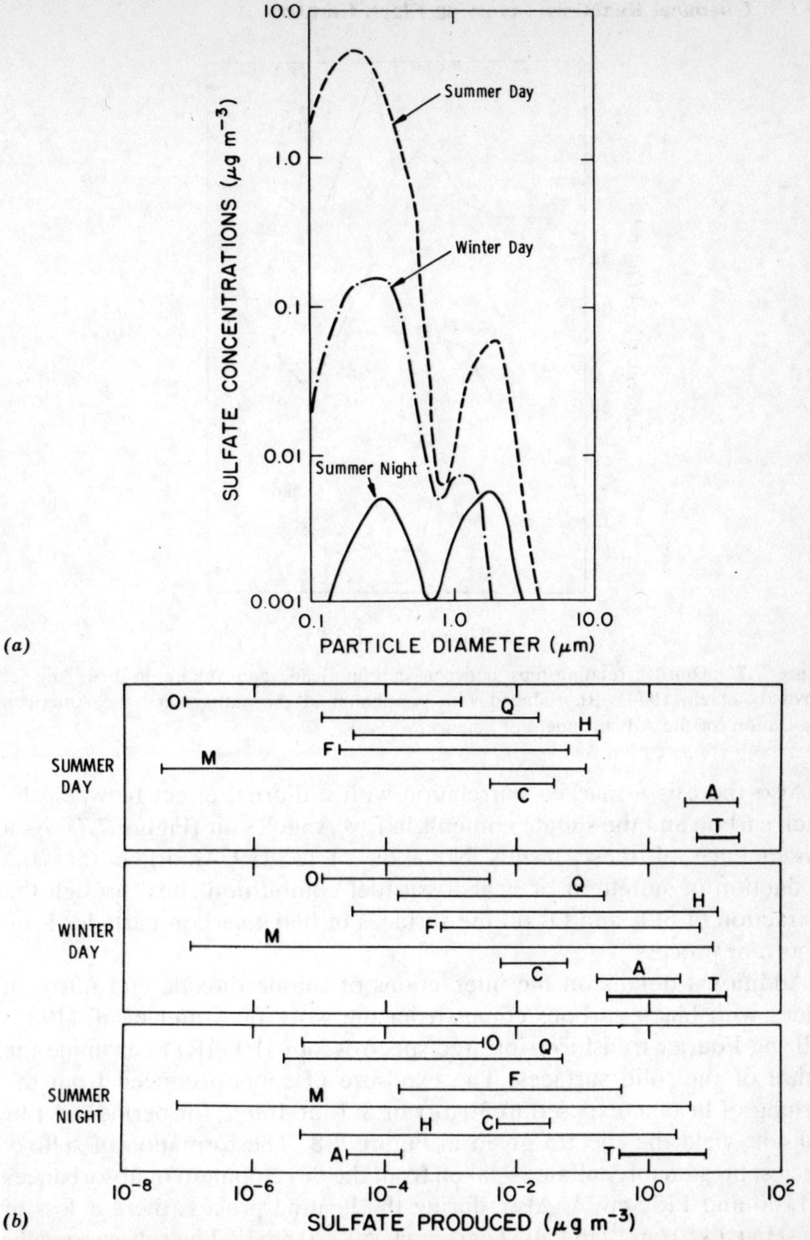

Figure 7.6. (*a*) Sulfate concentrations produced after 5 min under different atmospheric conditions in models (Middleton et al., 1982). Reproduced with permission of the authors. (*b*) Contributions of various mechanisms to sulfate production based on models of Middleton et al. (1982). T = total sulfate; $A = H_2SO_4$ condensation; $H = H_2O_2$ oxidation; O = uncatalyzed oxygen oxidation; M = manganese catalyzed oxidation; and C = black carbon catalyzed oxidation. Reproduced with permission of the authors and the American Geophysical Union.

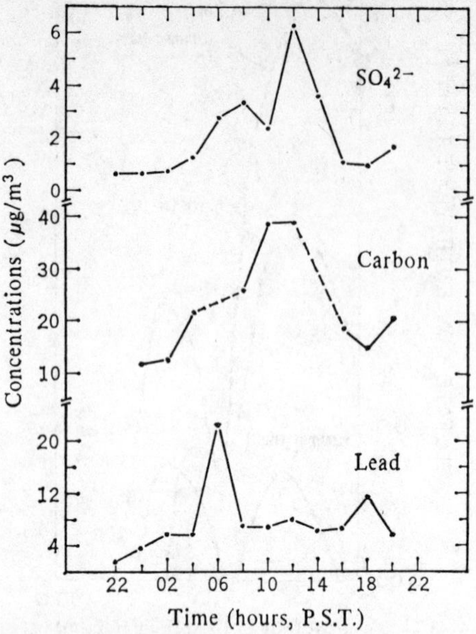

Figure 7.7. Diurnal relationships between carbon, lead, and sulfate in Los Angeles (Novakov et al., 1974). Reproduced with permission of the authors and the American Association for the Advancement of Science.

Also there is a marked correlation with a diurnal effect between the black carbon and the sulfate contents in Los Angeles air (Figure 7.7). As a consequence of these results Novakov et al. (1974) argue that the production of sulfate at or near fossil fuel combustion sites through the interaction of SO_2 and O_2 on the surfaces of black carbon particles is an important process.

Additional details on the interactions of sulfate dioxide and nitrogen oxides with black carbons comes from the work of Keifer et al. (1981) utilizing Fourier transform–infrared spectroscopy (FT–IR) to examine the nature of the solid surfaces. The exposure of soots produced from the burning of hexane exposed to 30 torr of SO_2 at 400°C for periods of 1 hr and 4 hr yield the spectra given in Figure 7.8. The formation of sulfoxy species, presumably sulfate, is taken from the development of absorbances at 1100 and 1135 cm^{-1}. Also, during the heating process there is loss of =C–H at 1440 cm^{-1} and alkyl carbonyl at 1720 cm^{-1}. The sulfoxy species increase with increased temperature and time of heating. They first appear at 300°C under these conditions.

The reaction of the hexane soot with NO_2 at a pressure of 30 torr after 70 hr and at room temperature is indicated by the FT–IR spectra shown in

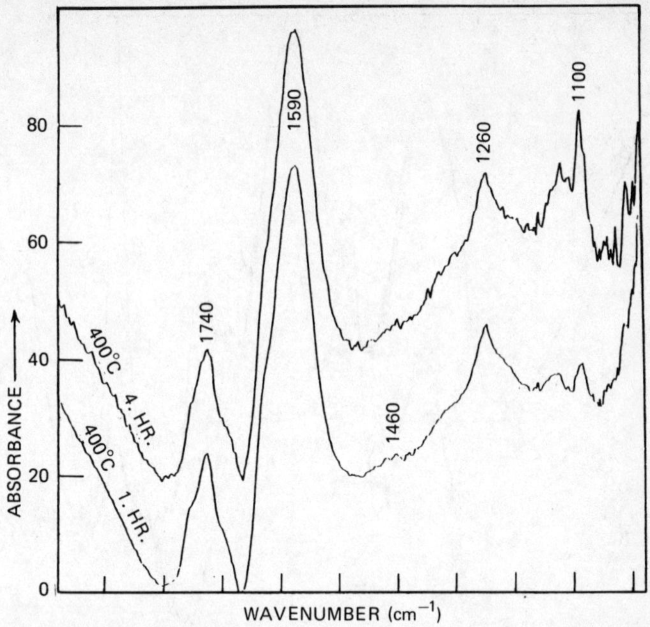

Figure 7.8. The IR spectra of the surface of soot exposed to SO_2 at 400°C (Keifer et al., 1981). Reproduced with permission of the authors and the Society of Photo-Optical Instrumentation Engineers.

Figure 7.9. The absorbances developed at 1660, 1540, 1340, and 1330 cm^{-1} herald the creation of new surface species. The 1540- and 1340-cm^{-1} bands are related to a surface $-NO_2$ species, as is evidenced by a shift of -35 cm^{-1} when the reactant used is $^{15}NO_2$ instead of $^{14}NO_2$. The 1660- and 1300-cm^{-1} absorbances derive from surface carboxylates.

The importance of these surface functional groups in the conversion of sulfite to sulfate was emphasized by Eatough et al. (1979) who studied the reaction in an aqueous solution of a commercial charcoal (Nuchar C-190N, from Bios Laboratory). The calorimetric experiments indicated that there are groups that are capable of complexing with sulfite and rapidly oxidizing sulfite to sulfate with the production of new reduced species on the charcoal. The rate limiting step for the oxidation of sulfite appears to be the oxidative regeneration of this active site.

The possibility of an increased production of diesel vehicles with a consequential increase in the emissions of soot and an increased catalytic oxidation of SO_2 has been put forth (Schryer et al., 1982). The well-known "urban sulfate anomaly," the lack of any covariance between decreases in atmospheric sulfates and decreasing fluxes of sulfate dioxide, have been

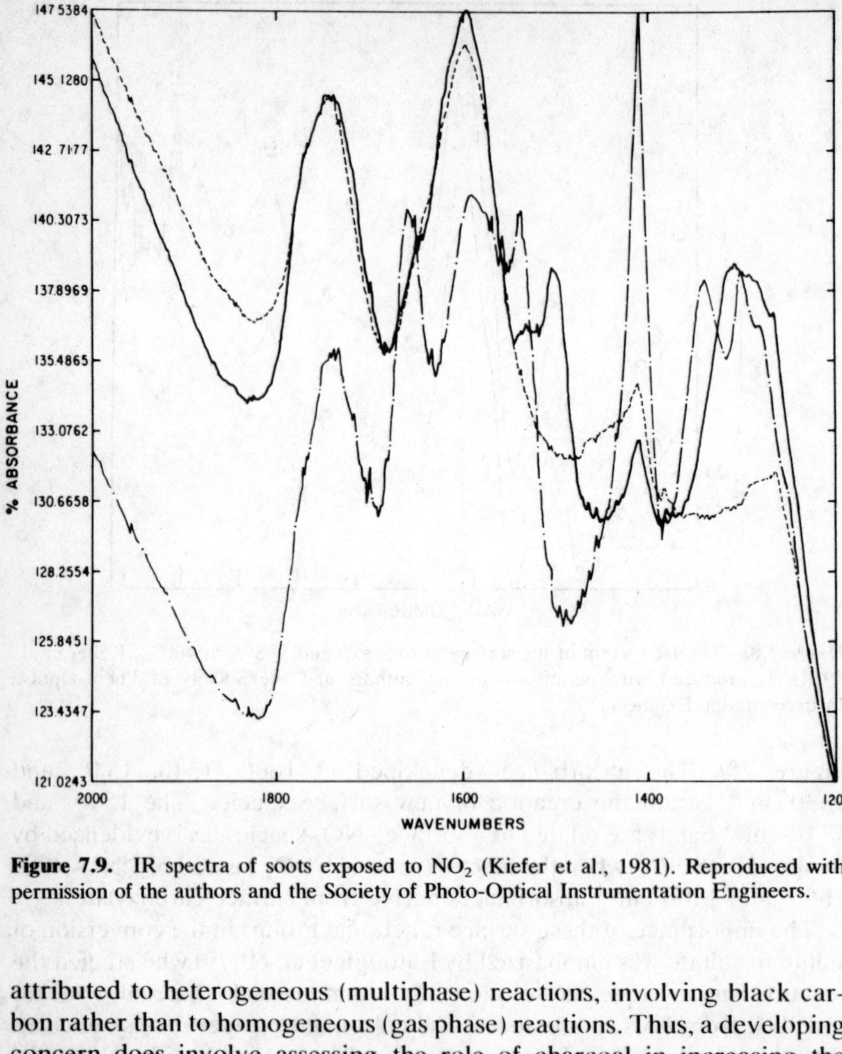

Figure 7.9. IR spectra of soots exposed to NO_2 (Kiefer et al., 1981). Reproduced with permission of the authors and the Society of Photo-Optical Instrumentation Engineers.

attributed to heterogeneous (multiphase) reactions, involving black carbon rather than to homogeneous (gas phase) reactions. Thus, a developing concern does involve assessing the role of charcoal in increasing the acidity of the atmosphere.

Calcite Conversion to Gypsum

The deterioration of marble surfaces exposed to urban atmosphere involves the transformation of calcite into gypsum (Del Monte et al., 1981; Del Monte et al., 1984b). This reaction appears to be readily catalyzed by black carbon cenospheres whose sources are oil burning. Coal produced particles have not been found to cause the transformations. Del Monte and his collaborators made photomicrographs of altered marbles from northern Italy and found spherical particles embedded in the sulfate–calcite altered layer. There is an usual association of the black carbon particles with altered calcite, while the unaltered calcite is free of the cenospheres. The investigators attribute the sulfur associated with these fly-ash particles to be the source of the sulfates. On the basis of their morphologies, most of the black carbon particles in the marbles appear to have been produced by oil burning.

Absorbed Polynuclear Aromatics

Polynuclear aromatic hydrocarbons have a remarkable resistance to degradation when adsorbed onto black carbon particles (Butler and Crossley, 1981). No reactions were evident for periods of up to 230 days when exposed to airs with or without 5 ppm of SO_2. However, nitration reactions did occur with exposure to airs containing 10 ppm of NO_x with the formation of nitro derivatives. The order of decreasing reactivity is anthranthrene, benzo(a)pyrene, benzo(ghi)perylene, benzo(a)anthracene, pyrene, benzo(e)pyrene, chrysene, fluoranthene, phenanthrene, and coronene. Curiously, the nitration reactions convert some of the carcinogenic polynuclear aromatics to noncarcinogenic nitro derivatives.

8

HISTORICAL RECORDS OF ENVIRONMENTAL BLACK CARBONS

INTRODUCTION

The historical records of burning may be found in marine and lacustrine sediments as well as in sedimentary rocks through the characteristics of the contained black carbons. One important facet of unraveling past events is the establishment of time, usually through paleontological, stratigraphic, or radiometric techniques. In this chapter we will consider the spectrum of investigations involving historic combustions that are revealed through the depositional record.

THE RECORDS

Marine Sedimentary Column

The recently demonstrated ability to recover continuous sections of marine sediments spanning the present to the Upper Cretaceous provides an entry to a consideration of natural burning processes over this period. Herring (1977) has investigated charcoals in the Deep-Sea Drilling Cores from the Pacific.

He examined the possibility that the present day latitudinal zonation of charcoal in the Pacific, found by Smith et al. (1973), might also prevail in the geological past provided that the materials persisted in the sediments without significant alteration. Then, the argument that charcoal fluxes to marine sediments are governed primarily by wind patterns and nontropical forest burning could be assessed through the Cenozoic.

Eleven sites in the North Pacific provided materials for this work, two

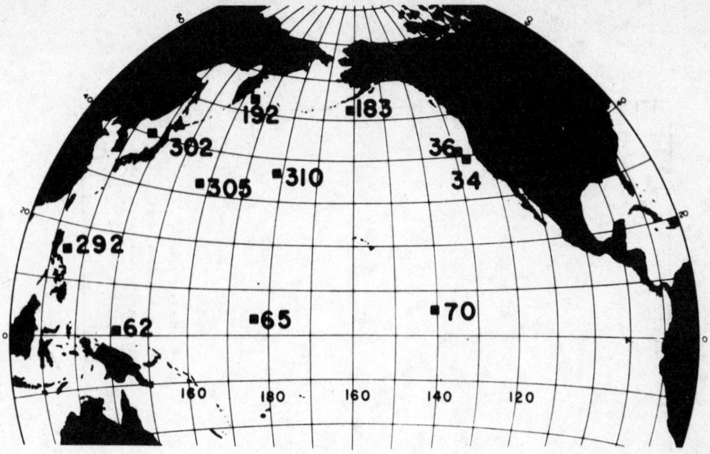

Figure 8.1. Site locations for deep-sea drilling cores for which charcoal concentration profiles were obtained (Herring, 1977). Figures 8.1–8.5 reproduced with permission of the author.

central ones and nine marginal ones. They are representative of low latitudes (Sites 292, 62, 65, and 70), midlatitudes (Sites 302, 305, 310, 34, and 36), and high latitudes (Sites 192 and 183) (Figure 8.1). The strata cover an age range of 65 million years. Cores were chosen on the basis that they contained a complete or nearly complete geological sequence and that ages could be placed upon them by paleontological techniques. In some cases, especially the work with the equatorial samples and the older samples, the charcoal contents were so low that only upper limits for the concentrations or fluxes could be given. Figures of charcoal fluxes to the deposit sites as a function of time utilize a shift in time scale to emphasize the past 10 million years.

Two examples from the data of Herring, representative of the North and Equatorial Pacific, will be considered. The first is from Site 192 in the Northwest Pacific (Figure 8.2). In this deposit there is a continuous decrease in charcoal fluxes falling from a present day value of a little less than 5 $\mu gC/cm^2/yr$ to about 0.02 $\mu gC/cm^2/yr$ at 37 million years ago. A similar pattern is found at Site 62 (Figure 8.3). The precision of the analyses is of the order of ±5%.

There is a persistence of carbon in the deposits for tens of millions of years. This can be seen in the fine details of the scanning electron microscope pictures (Figure 8.4). Herring points out that charcoals with detailed plant morphologies can be found in Oligocene sediments.

The distribution of charcoal concentrations in Quaternary sediments

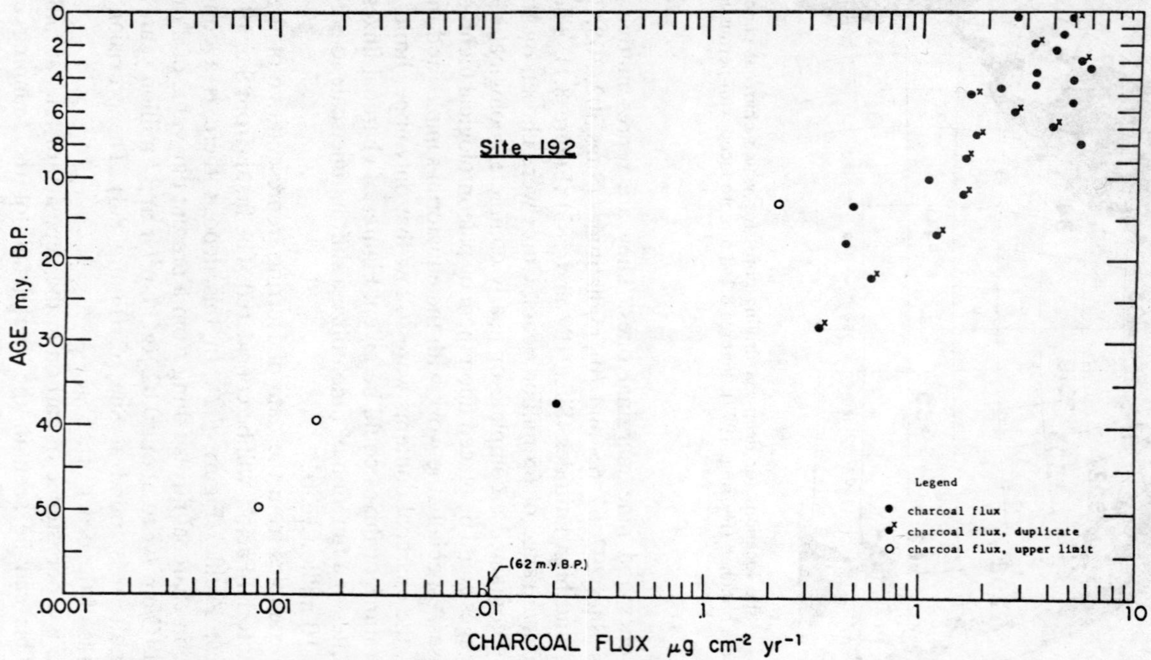

Figure 8.2. Charcoal fluxes to a Northwest Pacific sediment as a function of time (Herring, 1977).

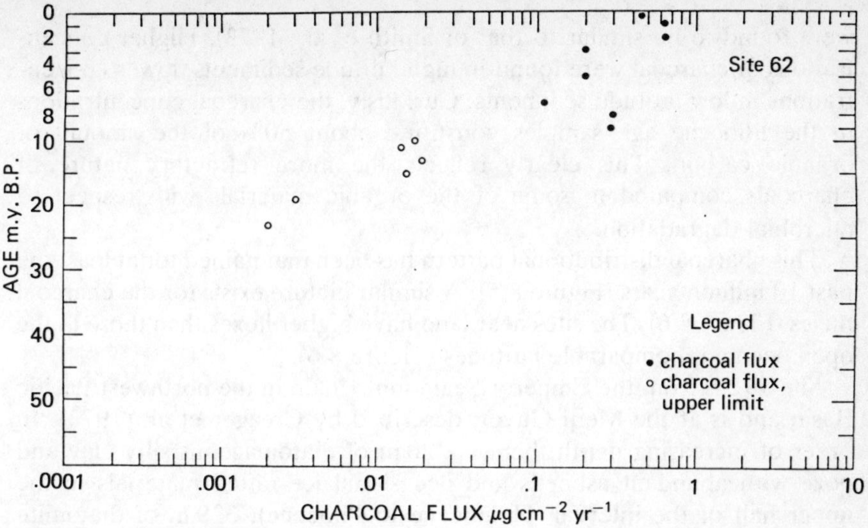

Figure 8.3. Charcoal flux to an equatorial region as a function of time (Herring, 1977).

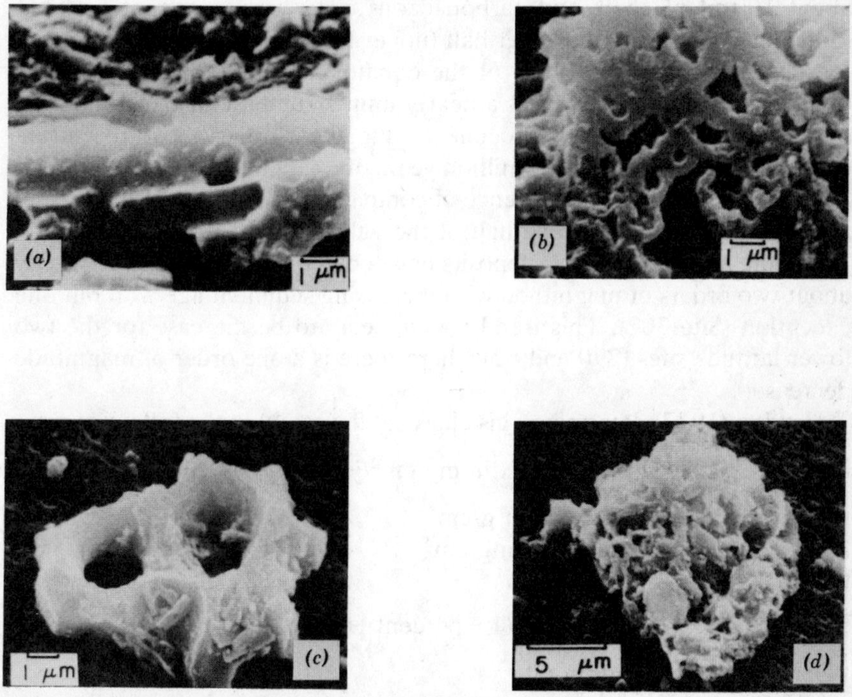

Figure 8.4. Scanning electron micrographs of charcoal particles from Site 62. Axial features in (*a*) are similar to tracheids or parenchmya in plant tissues (Herring, 1977).

were found to be similar to that of Smith et al. (1973). Higher concentrations of charcoal were found in high latitude sediments, lower concentrations in low latitude sediments. Curiously, the charcoal concentrations in the Pliocene age samples constitute about 50% of the amount of organic carbon. This clearly reflects the more refractory nature of charcoals compared to some of the organic materials with respect to microbial degradation.

This charcoal distributional pattern has been maintained for at least the past 10 million years (Figure 8.5). A similar picture exists for the charcoal fluxes (Figure 8.6). The sites near land have higher fluxes than those in the open ocean at comparable latitudes (Figure 8.6).

Site 192 is from the Emperor Seamount Chain in the northwest Pacific Basin and is at the Meiji Guyot, described by Creager et al. (1973). In order of increasing depth there is 320 m of diatomaceous silty clay and ooze with abundant ashbeds and occasional ice-rafted materials in the upper half of the interval (Holeocene to Pliocene); 329 m of diatomite ooze (Upper Miocene); 155 m of diatomite rich clay (Lower–Upper Micoene to upper middle Miocene); about 250 m of clay with minor calcareous horizons (lower middle Miocene to Oligocene); and finally about 100 ml of chalk and carbonaceous clays with minor intercalated sand and silt layers in the lower half (upper Eocene to upper Cretaceous).

Site 62 is located 2° north of the equator and 500-km north of New Guinea. The core sequence is a nearly uninterrupted chalk spanning the period from the upper Oligocene to the Quarternary. For the strata deposited during the past 15 million years, the sedimentation rate is about 25 m/million yr. As a consequence of compaction, the rate below this level is computed to be about one half of the value.

Middle and high latitude deposits have charcoal fluxes that decrease by about two orders of magnitude with increasing sediment age with but one exception (Site 302). This trend also appears to be the case for the two lower latitude sites (290 and 62) where there is a one order of magnitude decrease.

Herring (1977) determined his charcoal fluxes (Φ) in the following way:

$$\Phi = \text{mass flux in mg/cm}^2\text{/yr} = (S/10)\,D$$

where D = dry bulk density in g/cm^3
D_w = wet bulk density in g/cm^3

Thus

$$D = D_w\,(1 - \text{percent H}_2\text{O} \times 10)$$

where S = sedimentation rate
% H_2O = water content in percent

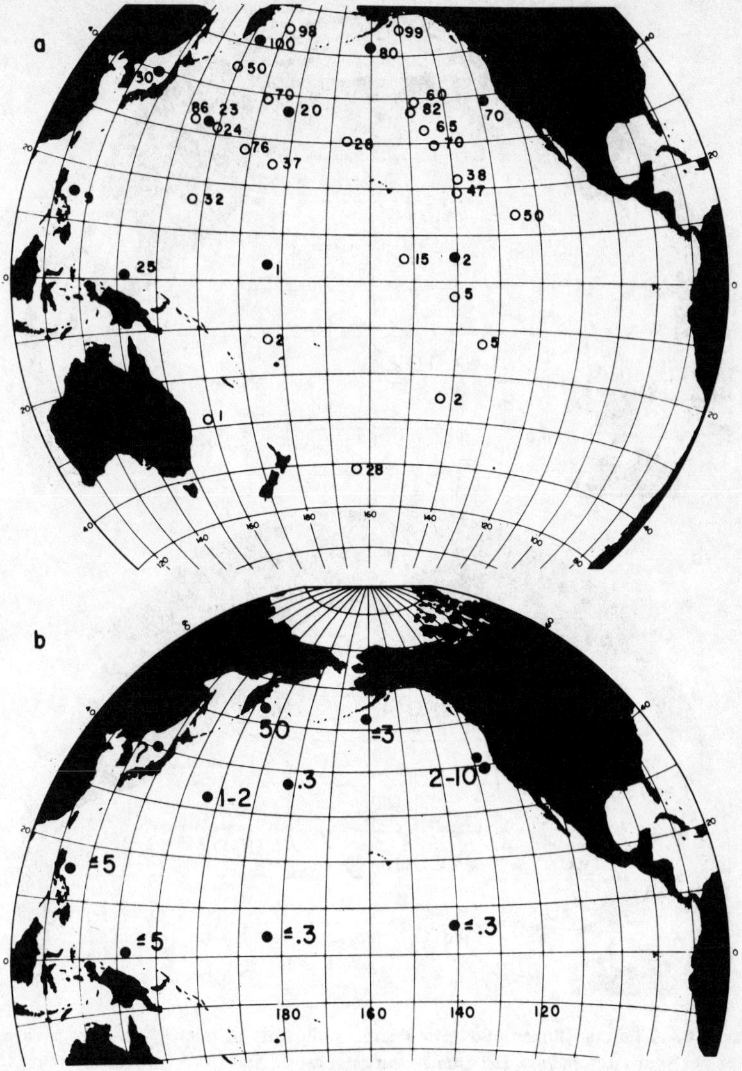

Figure 8.5. Charcoal concentrations in Pacific sediments in weight percent × 10^3 for the data of Herring (1977) ● and that of Smith et al. (1973) ○. (*a*) Quarternary averages, and (*b*) averages at 10 m.y. B.P.

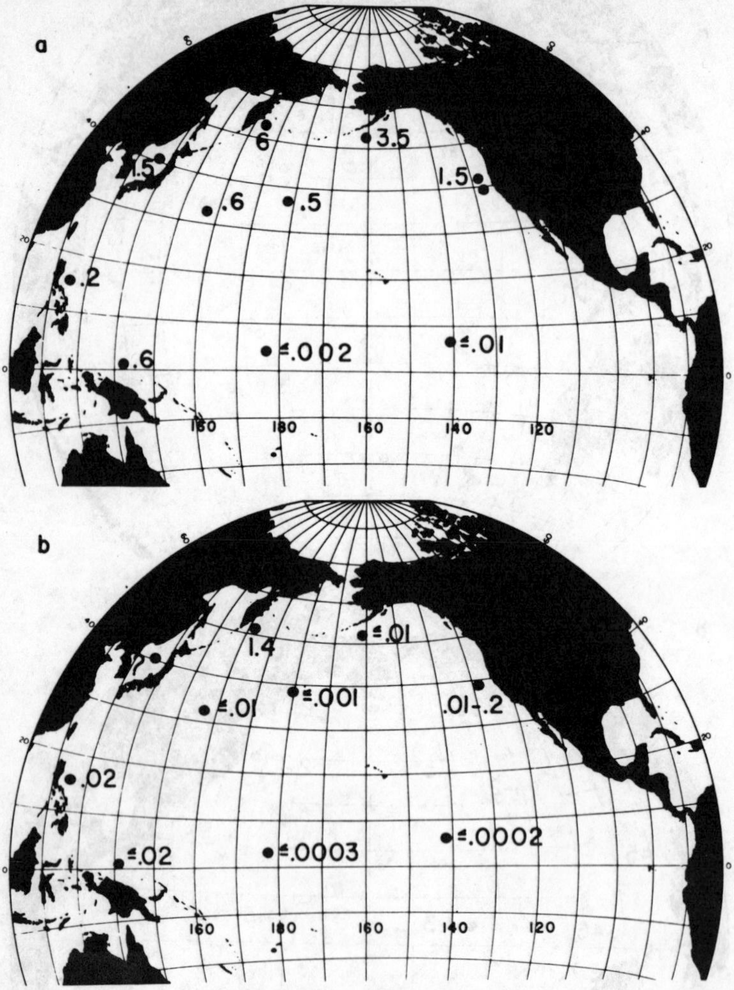

Figure 8.6. Charcoal fluxes of North Pacific sediments in micrograms per square centimeter per year ($\mu g/cm^2/yr$). (*a*) Quarternary averages, and (*b*) Averages at 10 m.y. B.P.

The following assumptions are used:

1. The density of water is 1.000 g/cm^3.
2. The partial molal volumes of the salt in the interstitial waters are negligible.
3. There is no unoccupied pore space.

In practice the water contents were determined by drying overnight at 110°C.

Sedimentary Rocks

A question that is relevant to many studies involves how well charcoal preserves the form of the original materials. The distortion of plant tissues upon carbonization has been poorly studied. Both lengthening and shortening of seeds has been observed by Stewart and Robertson (1971) in laboratory experiments. Grains were placed upon a hot pad and heated for 10 min. Those seeds with 11% moisture showed a shortening in length but a widening, between 6 and 12%. Grains with 15% moisture showed diminutions in both length and width. Distortions of up to 50% in size were noted. Clearly, the size of a seed at the time of carbonization may be quite different from that taken from the sedimentary record.

The occurrence of charcoal in the Newcastle sandstone, a Lower Cretaceous deposit found in Wyoming, Montana, and South Dakota, US, led Skolnick (1958) to develop a description of the area of deposition. The charcoal is abundant enough to impart a dark grey cast to the host rock. Its origin was most probably forest fires and it was carried by streams (and perhaps by winds) to the depositional area. The charcoal, floating on the transporting waters is probably carried very rapidly to the beach and bar depositional area. Following sedimentation, it would be out of reach of wave action and such a situation does explain its presence along cross-bedding planes. The presence of charcoal indicates the existence of a forested landmass, adjacent to the Newcastle area, during the lower Cretaceous. Further, he argues, lagoons bordering the seaward side through lensing of sand bodies provided an environment favorable for the entrapment of the charcoal.

The determination of the vegetation of an area for a given time period can often be determined from the charcoal record in sediments. The charcoal often is in an intact and noncompressed state. Harris (1957) studied the carboniferous flora of South Wales in materials taken from fissures of a limestone quarry. The fissures were filled with a grey calcareous clay, which showed little compositional variation, as did the flora, throughout the quarry. The charcoal remnants were recovered from a 1-nm mesh sieve, most of which came from the conifer *Cheirolepis muensteri*. The following categories of fossil were found among 306 fragments from a 200-g sample:

	Percent
Carbonized bark of Cheirolepis	51
Carbonized wood of Cheirolepis	28
Carbonized Cheirolepis leaves	7
Cutinized Cheirolepis leaf	6
Carbonized Cheirolepis microsporophyll	3
Carbonized Cheirolepis seed	1
Miscellaneous carbonized fragments	4
Total	100

Accompanying the plant charcoals were carbonized samples of beetles as well as some isolated resin bodies. The latter ranged from 0.2-mm spheres to spindles 0.7×0.3 mm. None were enclosed in tissues. The fissures had been invaded by recent root systems and small insects. Along with fossil pollens and spores were recent angiosperm pollen. Other flora associated with the conifer have been found in the German basal Lias and in the Greenland Rhaetic.

Harris associates the particles with a forest fire but does not consider atmospheric transport to the deposition site nor the incomplete combustion of biomass due to an insufficiency of oxygen. He argues that conifer parts, when burned in a bunsen flame, yield only a white ash. However, in a hearth fire, charcoal is found along with other delicate objects such as flowers, bracken pinnules, sporangia, and beetle fragments.

The Cretaceous fern, *Weichsella renticulata*, preserved in siltstone displayed remarkable internal structures as charcoal through electron microscopic examination. The association with partially charred materials is indicative of a high temperature origin. The fern in its charcoal form shows unique stomata and cells underlying the upper epidermis (Alvin, 1974).

Charcoalified fragments of plants offer a mode of preservation, often of very fine details, through much of the geological record. Scott (1974), for example, claims evidence on such a basis for the existence of the oldest Conifers (order Coniferales). He has isolated both compression fossils and charcoalified plant materials from thin clay bands in a coal. Single stoma were evident in the electron microscope pictures. The author points out that there is a growing evidence for a substantial amount of the flora being preserved in the charcoalified state.

Lacustrine Sediments

A 500-yr section, spanning the period 720–1270 A.D. in a laminated sediment from Greenleaf Lake, Algonquin Park, Ontario, Canada, was investigated using decadal segments for pollen, charcoal, aluminum, and vanadium analyses. Concurrent peaks in charcoal, aluminum and vanadium fluxes, varve thickness, and charcoal/pollen ratio were considered to be indicative of major conflagrations. Six fires appear to have occurred during the 500-yr period. The pollen analyses indicated a stable forest regime for the past 1200 yr. Thus, it is argued that fires are a natural part of the forest dynamics (Cwynar, 1978). The covariance of charcoal and vanadium and aluminum fluxes are indicative of the increased sedimentation and erosion consequential to the major fire periods. The charcoal fluxes provide the most direct evidence of fire. The varve thickness varies as a function of erosional activity with complementary increased aluminum and vanadium concentrations. Greenleaf Lake occurs along the Precambrian Shield. Sedimentary rocks are absent from the drainage basin. The forests are composed of a complex mixture of conifers and hardwoods. Cores were collected by the freezing tube technique. Charcoal analyses were made by the technique of Swain (1973). Aluminum and vanadium were determined by neutron activation analysis.

Several factors cloud the interpretation. First of all, the samples represent 10 yr of deposition although the large fires may have taken place over a period of but 1 yr. Also, there is the possibility that a varve was missed in counting. Occasionally there was a peak in charcoal influx not accompanied by increased aluminum and vanadium fluxes and varve thickness. Also varves may increase in Al and/or V concentrations and varve thickness with little increase in charcoal concentration or charcoal/pollen ratio. High erosion may be due to such a phenomenon as high precipitation.

Charcoals and Coals

Harris (1958) has observed charcoals in a variety of ancient environments, a phenomenon he attributes to forest fires. He emphasizes that he has not observed any charcoal produced by low temperature decay processes. He notes the differences between fossil woods preserved as charcoals and as coals:

1. Bituminous wood may exist as long tree branches grading down to leafy twigs, while charcoals from short angular or rounded lumps.
2. Bituminous wood is well compacted. Charcoals are friable.
3. Bituminous wood has a density of 0.75 while charcoal has one of 0.28.
4. Bituminous wood is compressed to the extent that its broken surface is glassy and usually showns no cells. Charcoals often possess cellular detail.
5. Bituminous woods evolve hydrocarbons upon heating. Charcoals have significantly less volatile material.
6. Bituminous woods are rapidly oxidized by $HNO_3/KClO_3$. They turn brown and in ammonia they nearly completely dissolve. Charcoal is more slowly oxidized and even after 2 months in the mixture only becomes dark brown and soft. In ammonia it yields tracheids with detailed structures in their pits.

The results are based upon charcoals from three areas: East Greenland at 70°N of basal Jurassic age; North Yorkshire at 54°N of Middle Jurassic age; and South Wales at 51½°N of Basal Liassic age. The plant materials are found in three categories: vertical roots in position of growth, charcoals, and bituminous coals.

The Oxygen Contents of Paleoatmospheres

The occurrence of charcoal in the geological record provides a lower limit for oxygen concentrations in the atmosphere (Cope and Chaloner, 1980). The argument is based upon the oxygen levels needed to combust carbon monoxide and methane. These are the two gases, released during the burning of plant materials, which support the flames. Data assembled by Cope and Chaloner indicates that the combustion of carbon monoxide will not take place in an atmosphere containing $<6.8\%$ oxgyen. Methane requires 12.8% oxygen in airs for its combustion. These values are 0.3 and 0.6 of the present atmospheric levels, respectively. Also, with decreasing oxygen concentrations, the combustion range for each gas becomes increasingly limited (Figure 8.7). Thus, the first appearances of charcoal in the geological column are indicative of atmospheric oxygen levels of at least 30% of the present value.

A review of the literature by Cope and Chaloner shows that the oldest

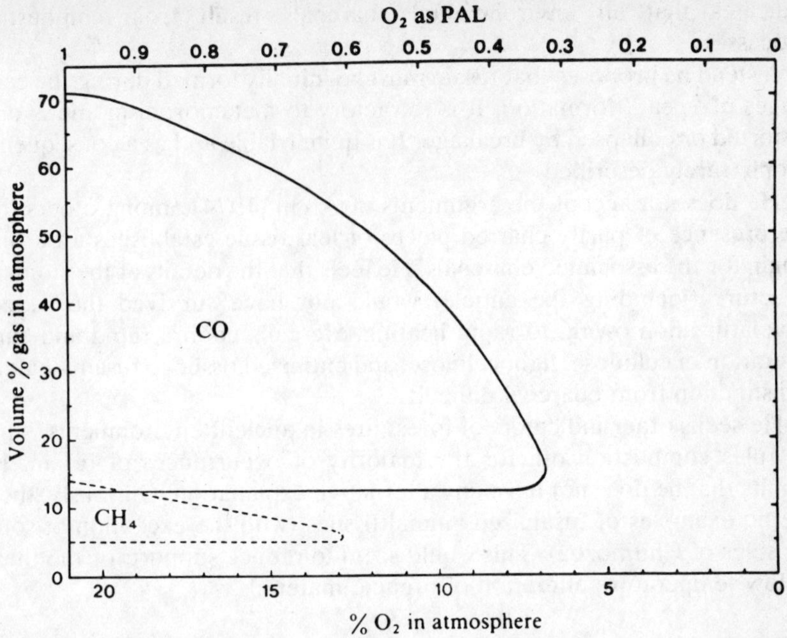

Figure 8.7. Combustion ranges of CO and CH_4 for varying percentages of oxygen in the atmosphere. PAL refers to present atmospheric level (Cope and Chaloner, 1980). Reproduced with permission of the authors.

sedimentary occurrences of charcoal in a noncarbonaceous matrix are found in lower carboniferous strata from Donegal, Ireland and Fife, Scotland. The oldest charcoals in carbonaceous matrices have been discovered in transitional Devonian-carboniferous coals from Pennsylvania. Thus, it can be induced that the oxygen levels in the atmosphere from the late Devonian (about 350 million years ago) attained values of at least 30% of that of the present.

Are Fossilized Charcoals a Product of High Temperature Combustion?

Schopf (1975) in a widely quoted article argues against a forest fire origin for the charcoals found in the sedimentary record even though they do bear a resemblance to burned materials. He bases his case on "its (fusain) widespread occurrence and lack of association with any evidence of conflagration." He does admit to a burning origin for some of the fusain in the environment, but he emphasizes that he does not consider that this indicates that all environmental charcoals result from combustion processes.

Instead he proposes that fusain must be initially formed during the early stages of "peat" formation. It is refractory to metamorphism and is only distorted or collapsed by breakage. It is quite friable and as a consequence is only rarely described.

He does not accept the arguments of Alvin (1974) among others that the presence of partly charred pieces of leaf tissue establishes a burning origin for the associated charcoals. He feels that the details of the stomatal structure (including the cuticle) would not have survived the sudden devolatilization owing to rapid heating. He calls upon a rapid and early alteration of cellulose, lignocelluose, and cutinized tissue to fusain, making a distinction from charcoal difficult.

He seeks other indicators of forest fires in ancient environments. Thus he rules combustion out for the majority of occurrences of fusain. He admits that he does not have any alternative explanation. Curiously, there are no examples of fusainized animal tissues (with the exception of some capsules of Chitinozoa). This would seem to reduce support for the thesis of low temperature alteration of organic materials.

THE INFLUENCE OF HUMAN SOCIETY

Society and Fire

Human societies have apparently husbanded fires for at least the past half-million years (Stewart, 1956) and have learned to produce fire since the last 10,000 to 20,000 yr. Stewart argues that a preoccupation with perpetuating fires existed during those early hundreds of thousands of years where the fires were produced by lightning, volcanism, or the spontaneous combustion of biomass. Clearly, the possibility that early societies caused extensive burning by careless fire practice seems high. Fire could also have been used extensively for the clearing of forests to make way for farming, the construction of homes, and for the hunting of animals after causing them to stampede. Fire provided warmth and perhaps kept predators away. It could be used to drive invaders from caves that early man wished to inhabit. It was used for cooking, fire heardening spears or tools. Further, fire could extend the length of day through the production of light. It may also have provided a basis for social activities through use in rituals or as a cohesive force in bringing people together.

The first evidence of fire making by human beings is attributed to an iron-pyrites ball with deep groove produced by striking to create a tinder igniting spark from a 15,000-yr-old Belgian site (Pheiffer, 1972).

The burning activities of Mesolithic man may be recorded in a peat profile in England through the deposition of charcoal and pollen (Simmons and Innes, 1981). Burning periods around 5000 to 6000 years ago are evident in the peat strata, which have been dated by radiocarbon assays, through high concentrations of charcoal. The particles are subangular, rather than rounded, which suggests they have not been transported over long distances. Whether there were burning intervals or aggregations of burning activities, is not evident. The pollen records indicate tree growth following the presumed burning episodes. The area is reputed to have one of the highest densities of Mesolithic flint sites in the United Kingdom.

Slash/Burn Practices

The aims of slash/burn practices in agriculture and forestry are the exposure of soil for further production of biomass, the reduction of fire hazards, and the elimination of unwanted species (Fritschen et al., 1970). Since complete combustion is not an objective of the activity, the introduction of charcoal to the atmosphere results. In general the lower the temperature of burning, the more incomplete it will be with the consequential production of larger amounts of charcoal (Fritschen et al., 1970). These investigators compared field tests with laboratory burns to ascertain the relationships between combustion characteristics and particulate emissions. The field studies were conducted with Douglas fir, western hemlock, and western red cedar in the State of Washington, USA. Broadcast fires represented low temperature fires whereas piles and laboratory fires simulated high temperature ones.

In the broadcast burns, ground levels of particulates increased to 10 times the ambient levels immediately downwind from the combustion site. The CO/CO_2 ratios increased from 0.034 during the initial stages of burning to 0.051 at an isolated hot spot. The gas samples contained ethylene, ethane, propene, propane, methanol, and ethanol, as well as other organics at much lower concentrations.

Laboratory fires had similar CO/CO_2 ratios ranging from 0.030 to 0.042 suggesting similarities in the combustion process. The importance of prevailing meteorological conditions is indicated from these studies.

Forest Fire History

The history of forest fires can be developed by the study of combustion scars on trees. The technique involves the identification of these scars and their positioning such that the year of occurrence can be determined by the counting of tree rings. The age of the fire is thus determined by counting the number of growth rings put down following the fire episode. The validity of the analysis depends upon a number of factors including an accurate determination of the ring count and the absence of damage of the record by insects. A curving of the ring often follows the fire insult and can be used for a more effective identification of the impact. The occurrence of scars on adjacent trees gives additional credence to the technique. It is interesting to note that trees can withstand repeated burnings and some have records of many fires.

The recent fire history of Barron Township, Algonquin Park, Canada has been studied through burn scars on living trees and their ages as determined by the number of growth rings that have formed since the fire episode. The period of 1939–1974 was investigated and there were about 0.19 fires/yr/100,000 ha. During the period 1696–1920 there were 16 fires in 225 yr, thus there is an average period of 14.1 yr between fires (Cwynar, 1977). In the recently studied periods, the average time to burn the entire area is about 70 yr. The largest fire occurred in 1875, following a drought season which confirms the sense that several fire years correspond to a period of drought. Forests depend upon fire for their maintenance and regeneration. The present vegetation appears as a consequence of the last fire. The origin of the fires (Cwynar, 1977) is given below:

		Man-Caused Fires 1939–1974				
	Lightning	Railway	Campers	Other	Total	Unknown or Miscellaneous
No. of fires	158	90	47	25	162	6
Percent of total	48.5	27.6	14.2	7.7	49.7	1.8

Fire destruction is usually followed by complete regeneration (Jones, 1945).

The frequency and intensity of fire in the forests of northeastern Minnesota is recorded in the sediments of the Lake of Clouds through their charcoal concentrations (Swain, 1973). Varve counting of the deposit allowed a time frame to be introduced that covered the past thousand years. The charcoal fluxes in terms of 10^6 $\mu^2/cm^2/yr$ are shown in Figure 8.8. The period 1660–1970 utilizes samples covering a 10-yr period, while the remainder of the samples span either 10- or 20-yr periods.

There appears to be a covariance between varve thickness and charcoal concentration below depths of 70 yr, according to Swain. This may be explained by the increased erosion following a fire on the slopes around the lake with the consequential increase in sedimentation. As the area becomes revegetated, erosion is reduced.

Coupling varve thickness increases and charcoal peaks lead to the following periods of local fires (Figure 8.8): 1680–1750; a series of at least 3 fires; 1590; 1520; 1390; 1300; 1170; 1020; and 980. Several charcoal maxima are not accompanied by increases in varve thickness (1790, 1460, and 1100) and these are indicated by "?" in the figure. The 1790 peak may be associated with known fires between 1801 and 1803. In 1240 there is an increase in varve thickness unaccompanied by an increase in charcoal. Finally, known fires are indicated by dates given in Figure 8.8.

Swain (1973) indicates that there have been around 17 fires over the past 1000 yr either recorded in the charcoals, fire scars, tree ages, or known events. The average frequency appears to be between 60 and 70 yr with a range of 20 to 100 yr. The charcoal fluxes to the sediments are somewhat lower after 1550 than before. Swain attributes this to the lower frequency of regional fires during the cooler, moister climates of the Little Ice Age of 1550–1850.

Finally, Swain argues that since several fires of the last 500 yr are not identifiable in the charcoal/varve thickness records, but are recorded in tree-ring analyses, the estimate of fires from charcoal/varve analyses are probably conservative. This in part probably results from a smearing of the records through the use of 10-yr increments in the sediment samples.

Forest Fire History in the United States

History of fire, both natural and anthropogenic, may be recorded in sediments. Otherwise, most histories are based upon anthropological evidence and inference. Human society presumably could utilize fire at entry into the United States (Butzer, 1971). This is based not only on

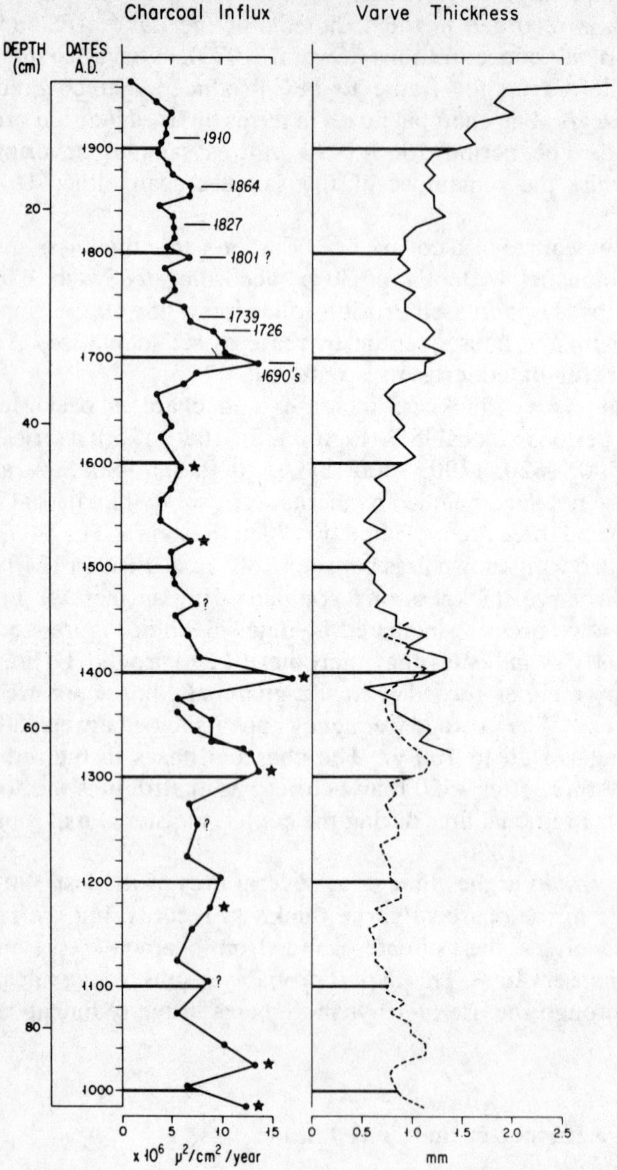

Figure 8.8. Charcoal fluxes to and varve thickness in sediments at the Lake of Clouds (Swain, 1973). Reproduced with permission of the author and of *Quarternary Research*.

campfire remains, but also upon evidences of extended grass and forest fires.

White (1972) points out that society can make a variety of adjustments to fire processes. It may alter the *source of fires* by setting new ones or suppressing those that were started naturally. People may *modify the hazard* of fires by changing the susceptibility of forested areas such as by the introduction of fire breaks or, finally, may *modify losses* by fire control measures such as improved systems of detection and firefighting. All in all, society does markedly influence the fluxes of carbon to the environment. White (1972) does indicate that intentional fires are the primary adjustment. They provide heat for cooking, for the preparation of tools, for protection from insects and animals, and for signaling. Broadcast burning of biomass aids in hunting by the driving out of game and in the preparation of land for cultivation.

White presents the annual numbers of fires in the United States and Canada (Figure 8.9) and from them it is evident that society is a significant force compared to lightning in the control of environmental fires. Differences in fire prevention, suppression, and protection programs of these two adjacent countries govern this data.

The flux of charcoal from biomass burning in the northwestern United States may be less today than it was in the period before colonization by the Europeans. Cramer (1974) considers three periods on the basis of the existing literature.

1. In the time before colonization he argues that there was much more burning than occurs at present. He points out that the entire ponderosa pine region burned every 4–18 yr from lightning strikes. The indigenous Indians used fire to clear growth and allow entry to west-side valleys.
2. This was followed by a period of increasing forest fire control measures, 1912–1940. As a consequence, there was a significantly smaller number of large forest fires. There were an increasing number of fires initiated by society. Natural fires were stopped as soon as possible after detection. Overall, the total annual extent of areal burnage decreased.
3. From 1940 to the present, there has been an accumulation of fuel in the forested areas as a consequence of the substitution of increased protection for hazard reduction. Lightning ignited large fires in such areas were large. The acreage combusted by wildfires may be increasing.

Cramer (1974) indicates that emissions of particulates and incompletely combusted gases to the atmosphere from biomass burning may be reduced by increased amounts of prescribed burning, especially that of accumulated forest fuels. As a consequence, there would be less danger of

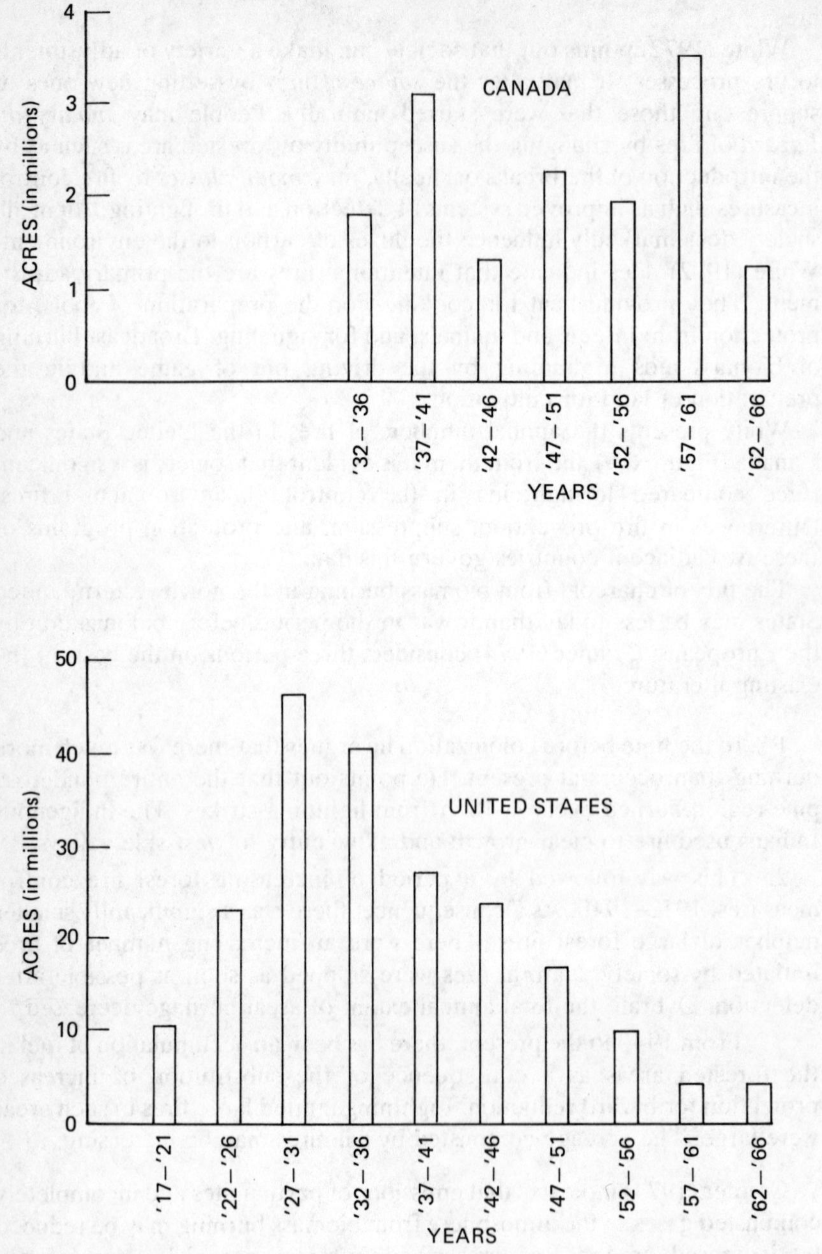

Figure 8.9. Fires in the United States and Canada (White, 1972). Reproduced with permission of the author.

large, destructive wildfires that can introduce large amounts of atmospheric pollutants. His literature survey indicates that charcoal fluxes will be significantly reduced by the higher temperature prescribed fires. Clean piles, properly prepared, burn with a higher efficiency and reduced fluxes of particulates, than do broadcast burns. For example, the particulate flux from a prescribed fire varies between 1 and 10 lb/ton, while the head of a wildfire releases about 58 lb/ton. Water contents also increase particulate emissions. Smoldering fires in litter and duff yield 3 to 10 times the particulate as do the flaming material. Grasses, when combusted produce 16 lb particulates/ton. Clearly, each fire will possess temperature of combustion depending upon the nature of the fuel, the nature of the ground material, and the atmospheric conditions.

There is a concern that the reduction of controlled forest burning, as well as the prevention and control of natural fires to maintain air quality, is creating serious hazards through the buildup of fuels in forested areas. The problem may be seen in Table 8.1 from Cramer (1974). There has been an increase in pile and burn activities but a decrease in broadcast fires. Perhaps of greater importance is the increase in forested areas receiving extra protection in lieu of treatment. Clearly, the possibility of large fires as a consequence is increasing.

Fire suppression may be the cause of the very large-scale combustion of southern California grassland, coastal sage scrub, and chaparral (Minnich, 1983). The buildup of herbaceous dead fuels is enhanced by the control of small fires. Consequently, small fires are replaced by large ones, difficult to control. Through satellite imagery, Minnich studied

Table 8.1. Activities in Slash Created and Slash Treated Areas of National Forests of the Pacific Northwest Region (in thousands of acres) (Cramer, 1974)[a]

Method	1963	1972
Clear cut	57	63
Partial cut	284	549
Broadcast burn	45	26
Pile and burn	0	87
Receiving extra protection		
Clear cut	0	36
Partial cut	512	912

[a] Reproduced with permission of the author.

controlled and all noncontrolled fires between 1972 and 1980 and noted the large burns may have occurred only since the beginning of suppression activities almost 70 yr ago.

Rain Forest Fires

Charcoal has been found in rain forest soils of the upper Rio Negro region of Venezuela in the north central Amazon Basin. There are occurrences in areas of known human settlement and in apparently uninhabited areas (Sanford et al., 1985). Carbon-14 dates of the soil charcoals span the last six millenia. The black carbons are most common in tierra firma forest Oxisols and Ultisols and less common in caatinga and igapo soils.

The occurrences predate the presently accepted times of initial human entry into the area. Thus, it is induced that climatic changes may have caused the fires that produced the charcoals.

This investigation destroys the assumption that lowland tropical forests have been free of fire disturbances. Dry weight concentrations of the charcoals in the soils ranged between 4.6 and 13.9 metric tons per hectare.

Forest Fire History in Panama

The role of a society in the production of environmental carbon is fashioned by both social and agricultural practices. These change with time. Often, historical records do not exist or are difficult to interpret. A melding of anthropological evidence, historical records, and sedimentary data may provide a picture of present and past practices that result in the entry of charcoal to the environment. For example, the first marked instance of charcoal in a Chilean lagoon sediment occurred at the same time for which there is the oldest evidence of human industry, 11,000 yr ago (Heusser, 1983).

Each society will be unique in involvements with fire and herein we will examine one to provide an insight into our present capabilities of understanding the past. We will consider the country of Panama and our information on anthropological evidence and historical records comes primarily from Bennett (1968) and Suman (1983).

Carbon-14 dates on artifacts (to a large extent charcoals) indicate that the first human entries to Central and South America occurred between 10,000 and 20,000 yr ago. The first stage of human activities, based upon a south central Colombia site, indicated that the people possessed fire. Their populations were apparently very small and their impact upon the ecology was apparently minimal. Following this culture was that of the

Paleo-Indian Stage (defined by Bennett) whose peoples were primarily hunters. The period of around 1500 to 8500 B.C. may have been covered. Further stages without any indication of agricultural practices included the Archaic Stage (6000 B.C.) where plant food may have been an important part of their diets. On the basis of estimates for South and Central America, Bennett suggests a range of possible populations from 6000 to 33,000. He further suggests that all of these cultures possessed fire which was used primarily in hunting and collecting. As a consequence, there were high probabilities of extensive vegetation burns. The widespread presence of fire resistant woody plants, argues Bennett, is indicative of burns during this period. He further states that before the advent of agriculture, fire had seriously altered the ecology of the Pacific side of Panama. Further, such a disturbance would have altered the fauna of the area through destruction of the flora necessary for their survival.

As yet, the determination of the time when agricultural practices began in Panama has not been made. One starting point is a time about 4000 B.C., but little is known about the agricultural activities. There is even a cloudiness with regard to the part of the isthmus that was first farmed.

The first Europeans visited Panama at the beginning of the 16th century. Bennett uses a very conservative estimate of perhaps 224,000 people in Panama at the time, with the greater part being in eastern Panama. The first accounts of the types of vegetation appear in the writings of Christopher Columbus, who made his observation in 1502. He noted well-populated regions with much cultivated land. Later writers indicated that fire was used to hunt deer.

There was a decided decrease in the Panamanian population following the entry of Spanish explorers as a consequence of disease, war, and employment of the Indians under harsh conditions. The first census (1932) suggests that the land that now constitutes Panama had slightly under 100,000 people. An 1851 census indicated 130,000. The important point is that from the time of the appearance of the Spaniards in 1500, there was a population decrease and most probably a decrease in slash/burn agricultural activities.

In the post Conquest period, shifting cultivation practices were dominant. The Spaniards introduced such crops as rice, bananas, and plantains to supplement the basic food crops of maize and tubers. Bennett indicates that fire was an easily and readily utilized resource. Much land was abandoned during this period and a period of reforestation occurred. Bennett suggests that throughout this period, there was an ecological retreat of society and the reestablishment of forested and wooded conditions over much of the isthmus.

The modern period, 1903 to the present, is characterized by an increase in human population and new agricultural practices. In the beginning of this period Panama became an independent republic and the United States began a period of canal completion and operation. The Panamanian census, taken in 1911, showed a population of 337,000 people that had by 1960 increased to somewhat over a million. There was a marked population growth in the agricultural areas of western Panama in which a doubling or tripling of populations took place. Half of the total labor force in 1960 was involved with agriculture. The shifting cultivation technique (roza) was employed in which brush and tree removal by steel cutting tools, such as the machete, and fire are important parts.

The annual burning seriously impacts upon the quality of the environment. Bennett points out that during the dry season, in late March and early April, the smoke is so dense that visibility from low flying planes is markedly reduced such that photography and reconnaissance are essentially not possible.

The relationship between charcoal production in slash/burn agriculture and the transport and deposition in coastal marine sediments has been investigated by Suman (1983) for the Gulf of Panama watershed. Burning in the western half of the watershed (12,000 km^2) yields 6.3×10^{11} g of charcoal annually based upon the equations of Seiler and Crutzen (1980) for the biomass burned for each vegetation type.

Charcoal fallout was monitored at six land stations during the burning period (dry season). Over 85% of the charcoal was in the fine fraction ($< 2\ \mu m$) and hence long-range atmospheric transport was possible. An annual deposition of 0.04×10^{10} g is estimated for that part of the watershed investigated.

Three box cores were taken from the sediments of the western portion of the Gulf of Panama to ascertain the input from this burning activity. The fluxes ranged between 84 and 469 $\mu gC/cm^2/yr$, values about an order of magnitude higher than those from the Santa Barbara Basin and Saanich Inlet sediments adjacent to the western coastal lands of North America (Griffin and Goldberg, 1975). This result emphasizes the high agricultural charcoal production in this area.

The majority of the charcoal is apparently carried by the rivers to the nearshore sedimentary sites. This is based on several observations. First of all, the flux of charcoal deposited on land was at least an order of magnitude less than that deposited in the nearshore sediments. Secondly, the fluxes for various size fractions were different for land and sea. On land, the charcoal fluxes decreased with decreasing particle size while in the marine sediments, they were uniform as a function of charcoal diameter. Continental runoff appears to be responsible for the large flux of fine particles to the coastal sediments.

Suman calculates that only a small percentage of the charcoal produced (63×10^{10} g) annually enters the atmosphere, 0.16 to 1.3×10^{10} g, while 0.08×10^{10} g C/yr enter the waters of the Gulf of Panama from this flux. A total charcoal flux of 3×10^{10} g/yr for the sediments of the western side of the Gulf of Panama is computed from a simple model.

The morphologies of the charcoal particles suggested that carbonized grass particles are more abundant than wood charcoals. In the sedimentary record there was no discernible shift in these charcoal types over the past two centuries suggesting a regular burning of the coastal savannas over this time period.

History of Fossil Fuel Burning

The morphologies, size distribution, and amount of black carbon in southern Lake Michigan sediments have provided a key to the history of atmospheric deposition from natural and fossil fuel burning as recorded in the sediments of Lake Michigan (Goldberg et al., 1981; Griffin and Goldberg, 1983). The concentrations of black carbon as a function of depth in the sediment or age of deposition of the strata as determined by Pb-210 assays are shown in Figure 8.10. The charcoal concentrations in the sediment prior to 1900 are extremely low and constant, usually under

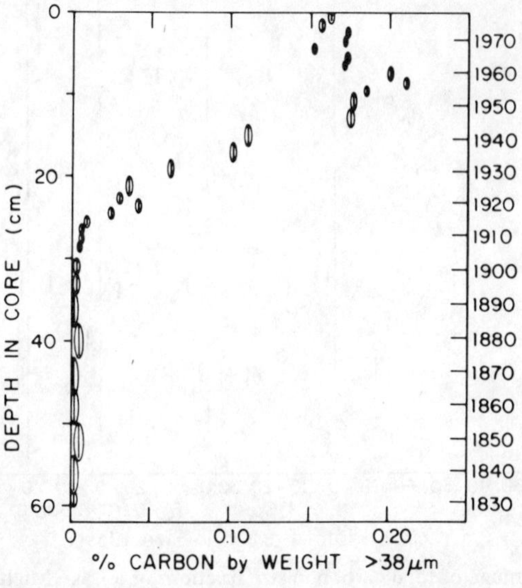

Figure 8.10. Black carbon concentrations as a function of depth (time) in a Lake Michigan sediment (Goldberg et al., 1981). Reproduced with permission of the authors and the American Chemical Society.

0.01% charcoal by weight. Beginning about 1900 there is a marked rise to 1960 and this is primarily attributed to the burning of coal. Subsequently, oil burning also contributed to these black carbon levels in the sediment. In the period 1953–1978, 76% of the carbon particles had an origin in coal burning, 14% in oil burning, and 10% in wood burning. The maximum is probably related to the installation of control devices at power plants during the 1960s and early 1970s. These efforts resulted in a significant decrease in the total suspended particulates in the atmosphere of Chicago, Illinois, which is at the southern tip of Lake Michigan.

The pre-1900 black carbons arose from wood burning, both natural and human-controlled. The fossil fuel burning then overwhelmed the inputs of black carbon to the sediments. The size distributions of the black carbons in these lake sediments amplify this picture (Griffin and Goldberg, 1983). Prior to 1900 the less than 1-μm fraction was dominant (Figure 8.11); the source was primarily biomass burning. In subsequent years the coal and oil burning became evident with larger particles, especially those >32 μm, making larger contributions to the total charcoal concentrations. The input of the larger particles is a consequence of their near fallout from energy producing plants near the lake. The

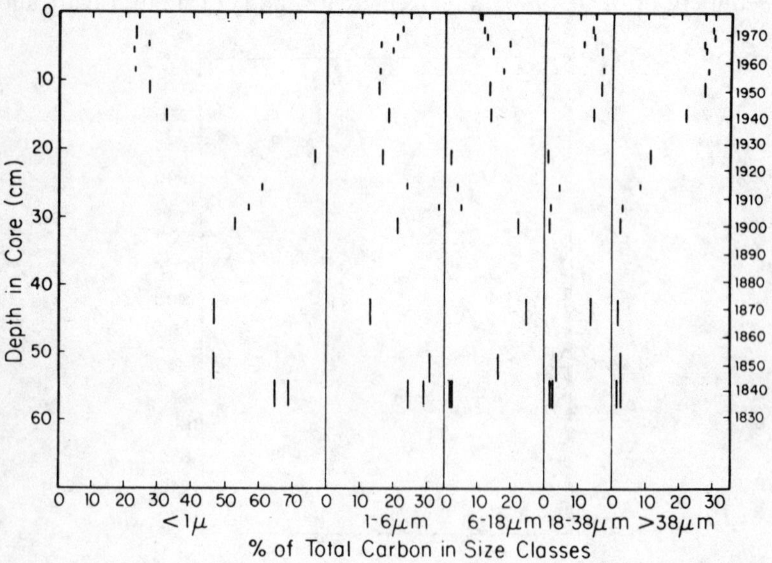

Figure 8.11. Percentage of total carbon in size fractions of a Lake Michigan sediment as a function of depth and the age of the strata. The geochronology was obtained by Pb-210 dating (Griffin and Goldberg, 1983). Reproduced with permission of the authors and the American Chemical Society.

fine-grained nature of the pre-1900 black carbons suggests more distant sources and consequential longer transport paths. Thus the sedimentary black carbons provide a picture of midcontinental United States burning where the forested areas were transformed into agricultural areas followed by the development of the present industrial economy.

Nine metals, Sn, Cr, Ni, Pb, Cu, Cd, Zn, and Fe displayed profiles similar to that of the black carbon particles (Figure 8.12), although the times of their maxima appear to be somewhat later than that of charcoal, perhaps, closer to 1968. Clearly, these metals are emitted to the environment from the fossil fuel burning and from industrial processes

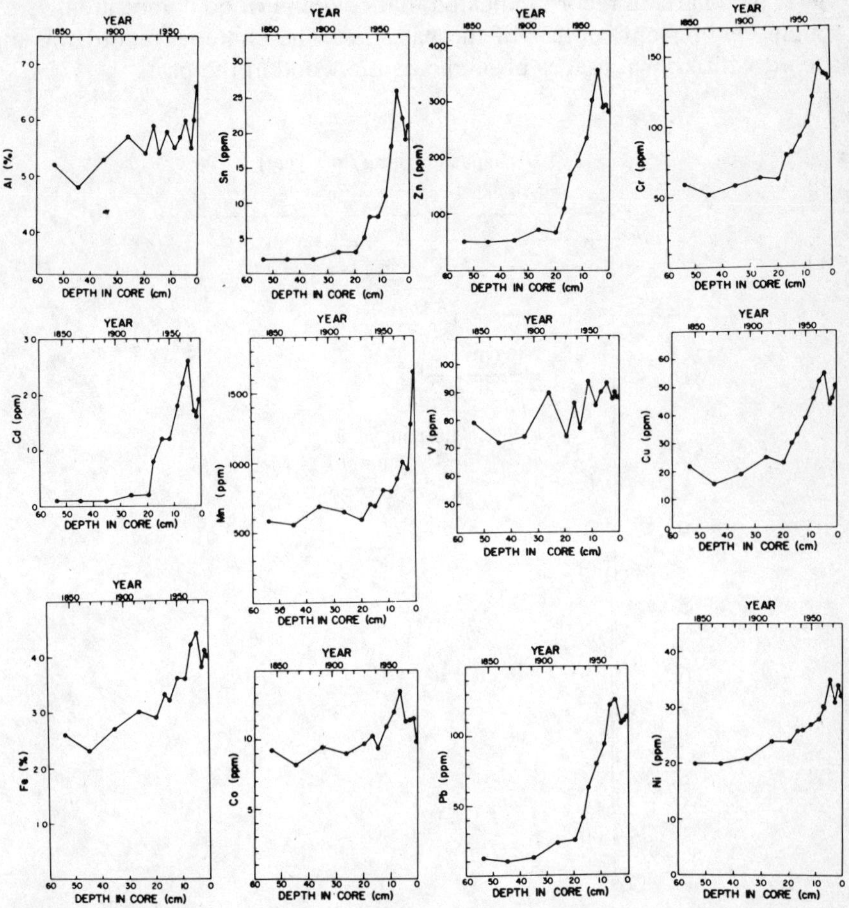

Figure 8.12. Metal concentrations as a function of depth and time in a Lake Michigan sediment (Goldberg et al., 1981). Reproduced with permission of the authors and the American Chemical Society.

associated with or dependent upon energy production. Three metals did not follow the black carbon. Manganese and Al displayed increases over the latter 20 yr of deposition, while V appears to have been slowly increasing since the 1940s.

The history of burning processes in an eastern area of the United States has been studied in a similar manner (Kothari and Wahlen, 1984). The concentrations and morphologies of charcoal particles greater than 38 μm in diameter, which were extracted from Green Lake, New York sediments, revealed that oil and coal burning were the dominant sources between 1905 and 1979. On the other hand, between 1590 and 1823 the black carbons originated from wood combustion. Between 1845 and 1961, the charcoal record indicated that coal and wood combustion were roughly equivalent sources of the particles. The charcoal record was in accord with known usages of oil, coal, and wood in the past.

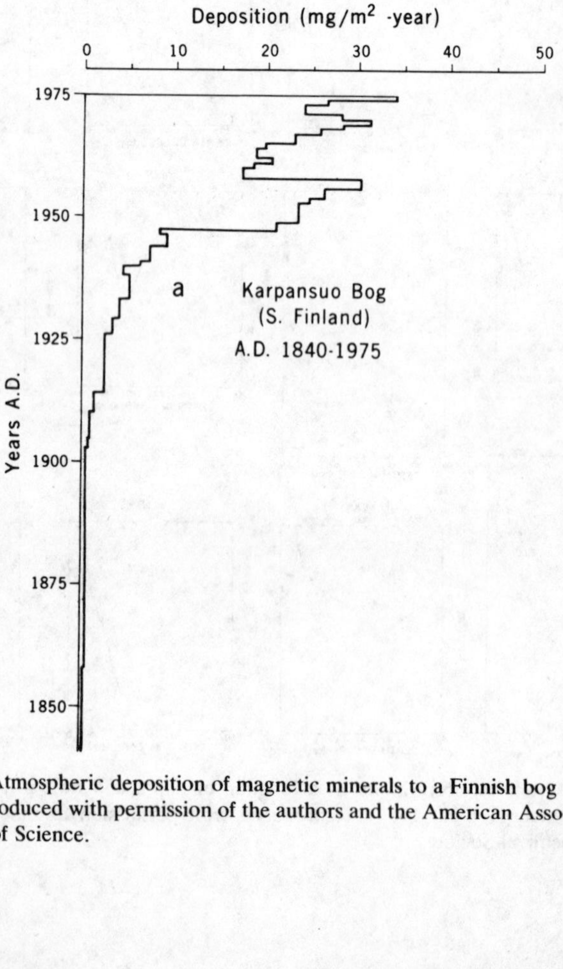

Figure 8.13. Atmospheric deposition of magnetic minerals to a Finnish bog (Thompson et al., 1980). Reproduced with permission of the authors and the American Association for the Advancement of Science.

Industrial activity on a global basis has been followed over the past century by investigations on the magnetic minerals in sediments, materials presumably in part introduced by anthropogenic activity, fossil fuel burning, and metallurgy. The substances, primarily magnetite, have environmental changes similar to those for the black carbons. For example, Thompson et al. (1980) measured the magnetic mineral concentrations in a Finnish peat bog where strata over the past century and a half could be dated. The resultant curve (Figure 8.13) is remarkably similar to the Lake Michigan charcoal profile.

History of Industrialization

The industrialization of New Haven, Connecticut, is recorded in the sediments of Lake Whitney, located 4 km from the central area of the city. Storm drainage empties directly into the Lake. There are no direct charges of either commercial or industrial wastes. The fluxes of black carbon to the sediments were determined by the technique of Smith et al. (1973) to ascertain the carbon concentrations in the strata which were dated by Pb-210 (Bertine and Mendeck, 1978):

Sample Depth (cm)	Element Carbon Flux ($mg/cm^2/yr$)
7–9	0.7
11–12	0.8
18–19	1.3
27–28	0.6
36–37	0.3
46–47	$\leq$0.1

The maximum flux occurred in the mid–late 1950s (18–19 cm) and decreased by 1964 (7–9 cm) perhaps as a consequence of emission controls being placed upon electric power plants. The scanning electron micrographs of the carbon particles isolated from the sediments resembled particles produced from coal, wood, and oil burning. The investigators suggested that the wood and coal derived particles could be differentiated from oil particles by their significant contents of Si, Fe, Al, and Cu.

History of Black Carbon in the Atmosphere

A historical data base for the loading of black carbon in the Los Angeles atmosphere has been developed for the past twenty-four years through recent analyses of archived high-volume air filters and filter tapes (Gray et al., 1984; Cass et al., 1984). Black carbon concentrations are determined either through analyses of the filters or through calibration of the reflectance of the tape samples.

The archived samples indicate that the black carbon concentrations in the Los Angeles airs have declined at most monitoring sites over the past 24 yr. In the late 1950s average daily black carbon concentrations occasionally exceeded 30 $\mu g/m^3$ whereas in the early 1980s the 24-h average values very rarely exceeded 20 $\mu g/m^3$. Over the 24-yr period the average black carbon concentrations ranged from a high of 6.4 $\mu g/m^3$ in downtown Los Angeles to 4.5 $\mu g/m^3$ in West Los Angeles. The black carbon concentrations, as well as these of the organic phases, show maxima in the winter months, a behavior characteristic of vehicle-derived pollutants such as CO and Pb. The ratio of total carbon to black carbon averaged over long time periods is 2.6 and shows little seasonal dependence. This observation suggests that the aerosol black carbon is local in origin and derives from vehicular traffic and nearby fuel burning.

9

THE IMPACTS OF COMBUSTION UPON THE ENVIRONMENT AS RECORDED BY BLACK CARBONS

INTRODUCTION

Present day burning activities, whether they involve fossil fuels or vegetation, are associated with a number of environmental problems. Black carbon acts as a tracer for the dissemination of particles about our surroundings from combustion. Some global problems have emerged as a consequence of burning and in this chapter we will consider the participation of the black carbons in their resolution. The role of black carbon in the global carbon budget, the long-range transport of combustion particulates, black carbons as carriers of carcinogens, and the impact of atmospheric carbons upon climate, will encompass some of the problems considered.

CHARCOAL AS A SINK IN THE GLOBAL CARBON BUDGET

The role of charcoal as a sink in the global carbon budget has been assessed by Seiler and Crutzen (1980). The following discussion is primarily based upon their work. Their arguments are based upon the total amount of biomass burned annually through natural and anthropogenic activities. The analysis utilized data gathered mostly between 1960 and

1975. They assume no degradation of charcoal with time through microbial or inorganic reactions. The following types of processes that lead to biomass combustion are considered:

1. Tropical shifting agriculture.
2. Deforestation.
3. Industrialization.
4. Natural or agricultural fires in Savanna areas.
5. Wildfires in temperate forests.
6. Prescribed fires.
7. Wildfires in boreal forests.
8. Industrial and domestic wood burning.
9. Burning of agricultural wastes.

The computations utilize the following model:

$$M = A \times B \times \alpha \times \beta$$

where M is the amount of biomass burned annually in a given biome, A is the area of the biome, B is the average amount of organic matter in the biome, α is the fraction of the average above ground biomass relative to the total biomass, and β is the burning efficiency. The more uncertain parameters are A and β. The presently available estimates of the parameters have been collected by Seiler and Crutzen (1980) and are given in Tables 9.1, 9.2, and 9.3.

This flux is estimated in another way. Charcoal represents about 40% of the particulates produced in temperate forest fires and in agricultural burning. Organic matter extractable into benzene constitutes about 50% of the weight while the remainder is mineral matter amounting to about 10%. Seiler and Crutzen point out that a most reasonable estimate of particulates from biomass burning is in the range of $(0.2 \text{ to } 0.45) \times 10^{15}$ g/yr with a consequential production of $(0.1\text{–}0.2) \times 10^{15}$ g of elemental carbon.

Another approach can be made utilizing the data of Griffin and Goldberg (1975) who provided some preliminary fluxes of black carbon to open ocean and coastal environments: for the former, based upon Pacific Ocean data, an entry of 60×10^{-9} g/cm^2 of surface/yr, while for the latter, based upon two coastal Pacific Ocean zones, a value of 50×10^{-6} g/cm^2/yr. Taking an area of 2×10^{18} cm^2 of land area and adjacent coastal water, we arrive at an annual black carbon influx of 0.1×10^{15} g/yr.

Table 9.1. Areal Extent, Average Biomass, Net Primary Product (NPP), and Above-Ground Biomass of Some Main Types of Plant Communities (Dry Matter) (Seiler and Crutzen, 1980)[a,b]

	Area (10^6 km^2)	NPP (g m^2/yr)	Phytomass (kg/m^2)	Biomass (kg/m^2)	Branches (%)	Trunks (%)	Roots (%)	Dead Above-Ground Biomass (%)	Above-Ground Biomass (%)
Tropical rain forest	17	2200	40	41	8	72	19	1	81
Tropical seasonal forest	7.5	1600	35	37	8	68	19	6	81
Humid savanna	15	1200	10	11	11	54	29	6	71
Dry savanna		900	2	2	10	43	37	10	63
Deciduous and broad leaved temperature forest	7	1200	30	33	7	56	27	10	73
Coniferous temperate forest	5	1300	35	38	9	55	27	10	73
Shrubland	8.5	700	6	7	54		36	11	64
Grassland (Steppe)	9	600	2	2	16		64	20	36
Boreal forest	12	800	20	25	5	53	25	22	75
Cultivated land	14	650	1	1	63		27	10	73

[a]Reproduced with permission of the authors.
[b]Tables 9.1–9.3 reproduced with permission of D. Reidel Publishing Co.

Table 9.2. Above-Ground Biomass in Selected Fire-Affected Ecosystems (Dry Matter) (Seiler and Crutzen, 1980)[a]

Activities	Above-Ground Biomass (%)	Average Biomass (kg/m^2)
Burning due to shifting agriculture	81[b]	39[b]
	71[c]	11[c]
Deforestation due to population increase	81[b]	39[b]
	71[c]	11[c]
Wildfires in temperate forests	73	35
Burning in dry savanna	64	2
Prescribed fires	75	5
Wildfires in boreal forests	75	25
Cultivated land	73	1

[a] Reproduced with permission of the authors.
[b] Tropical rain forests and tropical seasonal forests.
[c] Humid savanna.

The open ocean flux to an area of 3×10^{18} cm^2 is really not important.

It is inescapable with these estimates that the charcoal component of sediments and soils is a significant sink in the carbon budget. On the basis of presently available data, the annual flux of charcoal to the environment appears to represent about 20% of the total anthropogenic release of carbon to the atmosphere.

BLACK CARBONS AS CARRIERS OF ATMOSPHERIC CARCINOGENS

Black carbons are perhaps the most important constituent of atmospheric aerosols in the northern hemisphere and their association with polynuclear aromatic hydrocarbons (PAH) that have been found to be carcinogens in laboratory animals may provide a basis to study their involvement in human cancers.

The long-range atmospheric transport of 20 polynuclear aromatics, some of which have been reported to be carcinogenic, is evident from the investigations of Lunde and Björseth (1977). Airs were filtered in southern Norway and their trajectories were determined on the basis of meteorological data. The polynuclear aromatics apparently originated in France and

Table 9.3. Summary of Data for the Annually Burned Area and Biomass (Seiler and Cruzten, 1980)[a,b,c]

Activity	Burned and/or Cleared Area	Total Biomass Cleared	Biomass Exposed to Fire	Annually Burned Biomass	Dead Below-Ground Biomass	Dead Unburned Above-Ground Biomass
Burning due to shifting agriculture	21–62 (41)	31–92 (62)	24–72 (48)	9–25 (17)	7–20 (14)	16–72 (44)
Deforestation due to population increase and colonization	8.8–15.1 (12.0)	20–33 (26.5)	16–25 (20.5)	5.5–8.8 (7.2)	4.0–8.0 (6.0)	10.5–16.0 (13.3)
Burning of savanna and bushland	(600)		12.2–23.8 (18)	4.8–19 (11.9)	8–16 (12)	2.4–4.8 (3.6)
Wildfires in temperate forests	3.0–5.0 (4.1)	10.5–17.5 (14.0)	7.7–12.8 (10.3)	1.5–2.6 (2.1)	2.8–4.7 (3.8)	6.2–10.2 (8.2)
Prescribed fires in temperate forests	2.0–3.0 (2.5)	1.2–1.8 (1.5)	0.3–0.5 (0.4)	0.1–0.2 (0.2)	0.6–0.9 (0.8)	0.2–0.3 (0.3)
Wild fires in boreal forests	1.0–1.5 (1.3)	2.5–3.8 (3.2)	1.8–2.7 (2.3)	0.4–0.6 (0.5)	0.7–1.1 (0.9)	1.4–2.1 (1.8)
Burning of industrial wood and fuel wood		31–32 (31.5)	11–12 (11.5)	10–11 (10.5)	5.5	1[d]
Burning of agricultural wastes			19–23 (21)	17–21 (19)	27–31 (29)	1.9–2.3 (2.1)
Total	630–690 (660)	130–250 (180)	92–172 (132)	48–88 (68)	56–87 (72)	40–109 (74)

[a]Reproduced with permission of the authors.
[b]Units in 100-Tg dry matter and millions of hectares; to convert dry matter to carbon multiply by 0.45.
[c]Data in brackets represent average values.
[d]Excluding wood used in long lasting structures.

England as well as from northern Norway. They achieved atmospheric concentrations of 32 $\mu g/1000\ m^3$ of air.

The absorption of PAHs on charcoal may provide a disease vector by transport through the upper respiratory tract into the bronchioles and alveoli of the lungs. Particles that may enter the human respiratory tract have aerodynamic diameters of <5 nm. It is therefore necessary to know both the size distribution and the amounts of aerosols to assess potential dangers. Thomas et al. (1968) has indicated that for one of the most potent carcinogens [benzo(α)pyrene], its concentration upon soots produced from a variety of fuel combustion processes was constant. Thus, a first approach to the problem may be found in a knowledge of the black carbon contents of atmospheres in industrial countries.

Pierce and Katz (1975, 1976) have studied the distribution of PAHs and their atmospheric oxidation products, polycyclic quinones (PQus) in the Toronto, Canada area. In most of their samples, at least 50% of the aerosol mass was associated with particles whose size was less than or equal to 3 nm. Highest concentrations of PAHs were found in urban areas and the lowest values were associated with rural areas. There were seasonal differences in the association of the organic particles with the different size class. Such investigations emphasize the potential roles of black carbons in carrying these toxins to humans.

BLACK CARBONS, WEATHER, AND CLIMATE

There is growing evidence that the tropospheric burden of black carbons, primarily from industrial activity, can affect climate through its impact upon the earth's radiation budget. Although the necessary data are yet to be gathered, there is still every evidence that black carbons may play a significant role in the control of weather in some areas. Two types of involvements have been identified: the ability of atmospheric carbon through absorption or scattering of radiation to warm or cool the earth's gaseous mantle; and the ability of black carbons to change the albedo of snow and hence influence the amount and the rate of melting. In addition, the black carbons can absorb other atmospheric constituents as well as acting as a catalyst in chemical reactions. A recent review of the subject has been given by Toon and Pollack (1980).

The ability of black carbon particles to absorb radiation, given off by the sun, or reflected from the earth's surface is largely influenced by the single scattering albedo, ω_0, which is the fraction of light scattered by a single particle. For most particles, the value of ω_0 is close to 0.9 to 1.0. However, black carbons, since they absorb a substantial amount of visible light, have values close to 0.5. A second important parameter is the optical depth, τ, the weighted product of the number of particles in a column of unit area and the cross-sectional area of a single particle. The product, $\omega_0\tau$, is the scattering optical depth, that is, light passing through a column of air with aerosol particles will be reduced due to scattering by the exponent of the scattering optical depth. Thus, urban atmospheres with high contents of black carbon will absorb substantial amounts of light, which will be converted to heat, with an absorption optical depth of $(1-\omega_0\tau)$.

Climate changes would be recognized by a change in the global mean temperature, which would be produced by the balance of the solar energy absorbed against the infrared radiation delivered to space by the components of the atmosphere and earth's surface.

Toon and Pollack (op. cit.) indicate that small temperature changes may be quite significant. The temperatures have varied by about 5°C between the ice age and the present and by about 1°C over the past 1000 yr. The importance of ω_0 on temperature can be seen from Figure 9.1. These investigators suggest that if there were no aerosols at all in the atmosphere, the earth would be about 1.5°C warmer. Using a "standard aerosol distribution," they indicate that if the number of particles were increased by a factor of 2, the earth would be 1.5°C cooler. However, where absorption becomes important, as with black carbon particles, a warming effect can take place (Figure 9.1). Generally, for tropospheric aerosols, the earth's surface will cool if the aerosols overall have a ω_0 near 0.85. For lower values, a warming effect will be expected. Because of the heterogeneous nature and distribution of aerosols today, climatologists

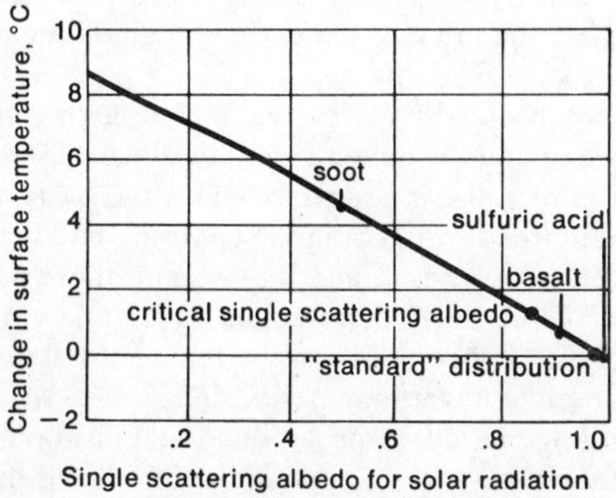

Figure 9.1. Single scattering albedo for solar radiation (Toon and Pollack, 1980). Reproduced with permission of the authors and *Sigma Xi*.

cannot stipulate whether there is a cooling or warming of the earth due to the presence of particles in the atmosphere. The optical properties of atmospheric aerosols, the sea salts, the soil debris, the sulfates, the charcoal, and so on, are quite different and the compositions of any given aerosol vary widely.

Whether the *anthropogenic* particles are warming or cooling the atmosphere is still undecided. The ω_0 of black carbon is such that if 10 to 20% of the atmospheric aerosols were composed of black carbon, a warming effect could take place. On the other hand, the sulfate aerosols, arising from fossil fuel burning, possess large values of ω_0 and, if abundant enough, could cause a cooling of the earth. Thus, restrictions of the release of black carbons, especially from diesel engines or of sulfur oxides from coal and oil burning will in principle have cooling and warming effects, respectively, upon the temperature of the atmosphere.

The areal extent of snow and ice has a considerable influence on weather and climate. The largest changes occur over periods of several days to months and in the northern hemisphere (Walsh, 1984). There is essentially no seasonal snow in the southern hemisphere. The amounts of snow and ice at any given time are determined to some extent by the melting, which is related to the albedos. For fresh snow the albedos can exceed 0.80 while for older and dirtier snows the values are around 0.5 to 0.6. The involvement of black carbon in changing the albedo of snows was first indicated by Warren and Wiscombe (1980). The carbons could cause an absorption of visible light resulting in melting.

A model to account for albedo changes has been formulated by Chylek et al. (1983) in which the black carbon particles are incorporated within the snow grains, as opposed to previous models in which the two materials were separated. Since the specific absorption of the carbon within the snow grain is higher than that of the carbon in air, less carbon is required to account for the observed effects.

A relationship between plant and tree fires in the northern hemisphere during the Holocene and the development of dunes has been proposed by Filion (1984). The hypothesis is developed on the basis of C-14 dates of charcoals and unburned wood fragments from the dunes from buried horizons. Between 5000 yr B.P. and the present, there are 13 maxima in the frequency of C-14 dates. The eolian cycle takes place during the cooling period of the Holocene. It is argued that natural fires initiate eolian activity and a subsequent warming period. The dunes would form and the subsequent warming period leads to dune stabilization. The following cooling period then allows for the initiation of fires and the completion of the cycle.

LONG-RANGE TRANSPORT OF COMBUSTION PRODUCTS

Black carbon particles can be used to study long-range atmospheric transport processes. Both natural and anthropogenic burning have provided the source materials.

The value of black carbon as a tracer for the atmospheric movements of sulfur oxides, acid ammonium sulfates, and trace metals from fossil fuel combustion was emphasized initially by Brosset (1976). Analyses of aerosols captured from Swedish atmospheres indicated that there were two classes of events or episodes characterized by the nature of the particles. Episodes were defined to occur when over 1000 particles/L of air were encountered.

The aerosols could be divided into four categories containing fine and coarse particles, under and over 2 to 2.5 μm in diameter, respectively, depending upon their chemical compositions and upon the wind trajectory that brought them to Sweden (Table 9.4). Black particles accompanied the aerosol Type I.3 (Table 9.5). The black episodes did not appear during the summer months but comprised two measurement periods, February–March and October–December. The black episodes were characterized by increases in the concentrations of manganese, iron, and vanadium. Their water soluble components consisted largely of ammonium sulfates and nitrates and the resulting solution is slightly acidic.

The white episodes have aerosols with small concentrations of black carbon and of nitrate in comparison to that of sulfate. The equivalent concentration of sulfate is matched by the sum of those of hydrogen ion and ammonium ion.

The black particles are brought to Sweden in trajectories from the southwest and east–southeast, that is, from continental Europe. These particles probably resulted from coal burning processes, although there is not a complete correspondence of high concentration of metals with increased concentrations of black particles (Table 9.6). The average compositions of the particles in the black and white episodes are given in Table 9.7.

Recently, there has been a great amount of interest in the Arctic haze whose components are transported even further than those in the Scandinavian studies. The sources for the pollutants are midlatitudinal fossil fuel combustion and industrial activity in the northern hemisphere. Transport takes place to a large extent in the prevailing westerlies. The size distribution of the particles is primarily in 0.1 to 0.5-μm class. The transport of the particles takes place within a mixing layer of below 1 km, that is, from source regions with about the same potential temperature as the deposition site (Joranger and Ottar, 1984).

Table 9.4. Mean Concentrations of Some Ions Expressed as Equivalent Fractions (%) in the Water-Soluble Part of Fine and Coarse Particles in Samples from Onsala, Sweden (March–May 1973) (Brosset, 1976)[a,b]

Aerosol	Trajectory	Fine Particles						Coarse Particles					
Type	Sector	SO_4^{2-}	NO_3^-	Cl^-	OH^-	H^+	NH_4^+	SO_4^{2-}	NO_3^-	Cl^-	OH^-	H^+	NH_4^+
I.1	NE–NNW	57	3	41	0	1	46	7	2	91	0	0	5
I.2	NNW–WSW	82	12	7	0	4	48	45	15	37	3	0	32
I.3	SW–ESE	75	22	4	0	8	69	29	36	27	8	0	21
II	WSW–SW	83	2	10	0	14	65	25	18	52	5	0	15

[a]Reproduced with permission of the author
[b]Tables 9.4–9.7 reproduced with permission of AMBIO.

Table 9.5. Observed Mean Concentrations of Fine and Coarse Particles and of Black Particles Expressed as $\mu g/m^3$ (February–May 1973) (Brosset, 1976)[a]

Aerosol Type	Fine Particles	Coarse Particles	Black Particles
I.1	13	15	2
I.2	7	31	2
I.3	29	12	16
II	48	24	3

[a] Reproduced with permission of the author.

The Arctic haze is of widespread occurrence with the strongest intensities in March, April, and to a lesser extent in August. Restriction in range visibility can be 5 to 8 km. Its composition is similar to that of many urban particulates with sulfates up to 2 $\mu g/m^3$, organic phases up to 1 $\mu g/m^3$, and black carbon from 0.3 to 0.5 $\mu g/m^3$. Profiles of black carbon up to 10 km revealed maximum concentrations at a few hundred meters (0.5–1.2 $\mu g/m^3$) compared to ground level (average of 0.249 $\mu g/m^3$) at Barrow, Alaska, and at various levels below 1 km in the Norwegian Arctic (Rosen and Hansen, 1984 and Hansen and Rosen, 1984).

Black carbon has been used as a tracer for the fossil fuel combustion products generated in central Europe and transported to the Scandinavian Arctic regions by winds. Some studies have arisen from concerns about acid precipitation in Scandinavia (Rahn et al., 1982). The aerosols contain not only hydrogen, sulfate, and nitrate ions, but also the metals Fe, Mn, V, Cd, and so on. In the initial studies, the charcoal was measured semiquantitatively through the color alterations of the filters by reflectance techniques. Such measurements did not consider the other constituents of the aerosol as potential interferences. Nevertheless, the work did allow two types of episodes to be distinguished much as in Sweden. The first, the so called white episodes, occurred during the summer in which little charcoal accompanied the high concentrations of sulfate and hydrogen ion, while the second were characterized as the "black episodes," when large amounts of charcoal were found in the aerosols. The winter events were distinguished by a covariance of charcoal and sulfate and by high concentrations of metals. The materials in the "white episodes" were related to their air transport from over the North Sea to Sweden, while the trajectories of the particles in the black episodes were not defined initially.

Table 9.6. The Concentration of Soot in Black Particles, as Well as H^+, NH_4^+, SO_4^{2-}, Mn, Fe, and V in Samples Collected During the Period February–March 1975 (Brosset, 1976)[a]

Date	Concentration of Black Particles ($\mu g/m^3$)	Concentration in $nmol/m^3$ of			Concentration in ng/m^3 of			Aerosol Type According to Trajectories
		H^+	NH_4^+	SO_4^{2-}	Mn	Fe	V	
February 11	3.5	112	136	122	15	<10	12	I.1–II
12	45.3	13	590	290	550	710	35	I.2–I.3
13	16.2	21	176	107	153	<10	46	I.3–I.1
14	~0	0.3	23	77	20	<10	9	I.1
March 1	2.0	1	280	140	19	<10	11	I.1–I.2
2	19.7	21	680	300	148	380	24	I.3
3	6.0	2	260	77	14	250	11	I.2
4	26.6	12	590	200	145	210	14	I.3
5	23.4	13	580	176	87	300	18	I.3
6	3.5	0.3	108	51	19	60	10	I.2
7	14.2	4	440	136	142	70	13	I.2–I.3
8	7.8	3	172	61	43	<10	9	I.2
9	6.3	3	220	75	32	<10	5	I.2–I.3
20	~0	Alk	12	10	1	<10	3	I.1
21	~0	0.2	23	15	10	<10	2	I.1
22	1.3	1.4	27	22	16	<10	2	I.1–I.3
23	2.1	Alk	56	27	<1	<10	5	I.1–I.3

[a]Reproduced with permission of the author.

Table 9.7. Comparison of Particle Composition During White and Black Episodes (Brosset, 1976)[a]

Episode Type	Concentration in equiv/m^3							Black Carbon ($\mu g/m^3$)
	SO_4^{2-}	NO_3^-	$SO_4^{2-} + NO_3^-$	NH_4^+	H^+	$NH_4^+ + H^+$	NH_4^+/H^+	
Black	308	33	341	342	7	349	51	23
White	317	<3	~320	233	75	309	3.1	2.3

[a] Reproduced with permission of the author.

Recent aerosols collected in the North American Arctic (Rahn et al., 1982) during the winter contain large amounts of black carbons. The air movements to the Arctic are controlled near the surface by the Icelandic low pressure and the Asiatic high-pressure systems. During the transport, changes in the chemical composition of the air masses, especially the pollutants, take place. Further, there is a mixing of air masses during the movement to the Arctic. Rahn and McCaffrey (1980) have studied such changes in the movement of air from Eurasia to Barrow, Alaska. As the air mass ages, there is a removal of particulates, the oxidation of SO_2, the wet and dry removal of SO_2, and a buildup and latter removal of Pb-210. Linear rate constants have been proposed by these investigators to account for the compositional changes. Verification of the pseudo-Lagragian model proposed by Rahn and McCaffrey (1980) was sought in the transport of polluted air masses from Europe to the Arctic.

Basic similarities between the black and white episodes were emphasized rather than the differences. Both events are alike in a number of ways. Quantitatively, the sulfate/black carbon ratios and the sulfate/Mn ratios for the eastern episodes only vary by a factor of 3 or less. The latter ratio has values that overlap between the two events. Further, on the basis of the rate constants, changes in such ratios, as the sulfate/V ratio, with time in a given air mass could be explained. The black and white episodes were found to occur during the same season, that is, spring, sometimes one right after the other. The air trajectories for such black/white events are not terribly different in length; both have origins in heavily polluted areas—central Europe for the black episodes and eastern Europe/western USSR for eastern white episodes. Each type of episode ends up with the same final sulfate concentration. The investigators suggest that white episodes may have derived from black episodes, where in the aging process, the heavy metals and the charcoal are somehow removed.

The process for the initiation of black and white events is considered to be the same. Both start with a particulate source from coal burning. The main difference at termination is the aging undergone by the air mass. In the beginning they have the same primary constituents, heavy metals, charcoal, and SO_2. With time, the sulfur dioxide is converted to sulfuric acid through oxidation, while the charcoal and heavy metals are removed from the atmosphere in dry or wet fallout. Thus, the subjective and qualitative darkness of an aerosol decreases steadily with time. The time periods are seen to vary by a factor of 2 with respect to the distances covered by the white and black events. The presumed source for the black episodes is central Europe which is about 700–800 km from Sweden. For the black episodes, a day's travel at 6–7 m/s is calculated. For the white eposodes, it takes 3 days to arrive from eastern source areas. On this basis,

Rahn et al. (1982) estimate under direct northward flow it takes 10 days to reach the Norwegian Arctic and 20 days to Point Barrow, Alaska. It should be emphasized that these times are still imperfectly known.

The origin and extent of Arctic haze, the condition of reduced visibility in the northern regions caused through light scattering by aerosols, have been considered through metal/black carbon ratios (Barrie et al., 1981). The black carbon to vanadium ratio may be indicative of atmospheric particulates produced in Siberia (low values as a consequence of coal burning) or in North America (high values as a consequence of oil burning). The intensity of the Arctic haze undergoes annual cycles reaching maxima sometime in March to April. The source areas were identified through the trajectories of air particles. The weather charts indicate Siberia and North America as the source regions for the particulates during December 1979 and January 1980, respectively, whereas European sources were dominant in the early Spring 1980. The meteorological and chemical evidences, that is, black carbon to vanadium ratios, are in concord.

The manganese and vanadium in aerosols, not derived from crustal rocks, appear to be diagnostic of the sources of aerosols in the Norwegian and North American Arctic in addition to evidences from meteorological observations (Rahn, 1981a,b). The Mn/V ratio is greater than unity in aerosols sampled in Eurasia and less than unity for those in the northeast United States. The low ratios are assumed to be influenced by the combustion of Venezuelan residual oils, high in vanadium porphyrins, in the United States. The aerosols of the Norwegian Arctic appear to have a source in Eurasia on this basis. The North American Arctic aerosols require an additional source within Eurasia, with a higher Mn/V ratio, to account for their compositions.

A more elaborate scheme to follow aerosols produced by fossil fuel burning involves seven elements that can characterize emissions from regions in the United States and northern Europe (Rahn and Lowenthal, 1984). The ratios of six elements (Sb, noncrustal Mn, In, Zn, noncrustal V, and As) to Se were used to characterize the sources of the aerosols. Se was used as a normalizing agent because "it is a general pollutant found at similar concentrations in diverse source areas and hence will not bias the ratios toward any particular region."

There are systematic variations in the black carbon size distributions from sources in Europe to the Arctic (Heintzenberg, 1982). In the European source regions, the Aitken nuclei contribute strongly to the aerosol volume found below 0.9 μm. In urban areas about 50% of the black carbon was found in the Aitken size range. The aerosols, reaching the Arctic region after several thousand miles of transport, are strongly depleted in the Aitken size range as a consequence of coagulation, condensation, and precipitation processes.

The black carbons in aerosols collected over Atlantic airs between Hamburg, Germany and Montevideo, and Pacific airs between Ecuador and Hawaii were shown to have origins in both fossil fuel burning and combustions of vegetation and forests in tropical regions (Andreae, 1983 and Andreae et al., 1984). The trajectories of the sampled air masses identified continental sources for the black carbons. The material that had its origin in tropical burning processes showed a strong covariance with excess potassium, that is, the amount of the element in the aerosol that was not associated with sea salt or soil particles.

10

THE FLUXES OF BLACK CARBON TO THE ENVIRONMENT

A mass balance for the flow of black carbon about the environment cannot be made since the mechanisms for its destruction are yet to be determined. Still approximations to the fluxes, both natural and anthropogenic, can be sought. Present evidence indicates the values given in Table 10.1.

The commercial production of carbon blacks is primarily directed to use in automobile tires. In the US, which accounts for around 50% of the global production, over 90% of the carbon black is incorporated into tires. A part of this carbon black enters the environment through abrasion; a part through the destruction of the tire itself.

The emissions of black carbons from diesel and gasoline burning will probably increase in the future, not only as a consequence of increased numbers of vehicles but also because engines might be designed to be less efficient in the production of particulates, which are to a large extent black carbons, in order to reduce the emissions of NO_x (NAS, 1981). There is a concurrent increase in hydrocarbon and particulate emissions, as well as in fuel consumption, with decreasing emissions of NO_x for alterations of controls on present day engines (Table 10.2).

There is an expectation that the present percentage of light-weight vehicles in the United States that are diesel (2%) will increase in the future (Cuddihy et al., 1984). Clearly, this will depend upon such factors as government regulations, the cost and availability of diesel fuel, and customer satisfaction. Still, it seems inevitable that increased black carbon emission will come from diesel engines in all types of vehicles (see Table 10.1).

Projections for the State of California have been made for effect of diesel vehicle black carbon emission on visibility for the 1990s based upon

Table 10.1. Global and United States Fluxes of Black Carbons and Graphite

Source	Flux in 10^{12} gC/year		Reference
	Global	United States	
Biomass burning			
Fireplaces		0.011	Muhlbaier and Williams (1982)
Agricultural wastes, savannas, forests (wild and prescribed) and fuels	500–1700		Seiler and Crutzen (1980)
Fossil Fuels			
Diesel engines	0.1–0.2		NRC (1982), p. 109
(light-weight vehicles)[a]		0.045[b]	Cuddihy et al. (1984)
(heavy-weight vehicles)		0.120	Cuddihy et al. (1984)
Gasoline autos		0.0035	Muhlbaier and Williams (1982)
		0.068[c]	Cuddihy et al. (1984)
Diesel autos		0.0042	Muhlbaier and Williams (1982)
Coal burning power plants		0.005	Fisher et al. (1979)
Gas furnaces, residential		0.000018	Muhlbaier and Williams (1982)
Gasoline engines (heavy-weight vehicles)		0.008[c]	Cuddihy et al. (1984)
Commercial Production			
Carbon blacks	3.5	1.5	*Minerals Yearbook* (1976)
Graphite mining and production	0.6 (production)	0.05 (consumption)	*Minerals Yearbook* (1982)

[a]Light-weight vehicles are usually defined as automobiles and trucks with gross weights <3.85 tons (Cuddihy et al., 1984).
[b]Extrapolated to future years when diesel represents 20% of the light-weight vehicles. Currently, they represent 10%; also, the value assumes that 76% of the total suspended solids are black carbon.
[c]For gasoline engines, 10% of total suspended particle emission is assumed to be black carbon.

Table 10.2. Engine Emissions (NAS, 1981)[a]

NO_x	Particulates	Hydrocarbons	Fuel Consumption Penalty (%)
8	0.4–0.5	0.6–0.8	0
6	0.5–0.7	0.7–1.4	2.5–4
4	0.6–1.0	0.8–1.7	7–12

[a]Units are grams/brake horsepower-hour.

1980 estimates (Trijonis, 1984). Heavy-duty diesel trucks apparently account for one half of the carbon emission in 1980 and for about 5–15% of the light extinction. Assuming no controls, visibility will decrease by 9 to 35% from 1980 to the early 1990s at which time heavy-duty trucks would be responsible for about three-quarters of the emissions of the entire diesel fleet.

APPENDIX: ANALYTICAL TECHNIQUES FOR BLACK CARBONS

INTRODUCTION

There are two classes of analyses for environmental black carbons: (1) those that are concerned solely with qualitatively identifying a particle as a black carbon; and (2) those that seek a quantitative assay. The sophistication of the required equipment varies widely from optical microscopes to cyclotrons for deuteron activation analysis.

The electron microscope, and to a lesser extent the optical microscope, have been most valuable in assigning origins to atmospheric and sedimentary carbons on the bases of their morphologies. Prior chemical treatments remove most of the other noncarbon particles from the samples.

Raman backscatter spectra can be used to uniquely identify black carbon particles. Laser beam illumination, introduced into a modified optical microscope may prove most useful in research investigations. Particles with sizes greater than several microns should be easily analyzed.

The quantitative analyses of black carbons involve a number of techniques, some of which may not give reliable results. For example, many ecologists and paleontologists in their sedimentary studies utilize the optical microscope for counting the numbers of particles above a certain size, usually with visible areas $>50\ \mu m^2$. Thus, all particles smaller than this will escape inclusion in the assay. Further, the visual identification of some particles as black carbon can be indecisive. In some sediments there are both what appears to be completely charred fragments and partially burned plant biomass. Some investigators include the latter with their estimate of charcoal abundance whereas others do not. One investigator only included charcoals with obvious cellular structures.

Infrared and Raman studies suffer from the complex nature of their spectra, a combination both of transmission and reflection. This is a consequence of the ability of black carbons to absorb infrared radiation throughout the entire spectrum. A technique to avoid this difficulty may rest in internal reflection infrared spectroscopy.

Some of the techniques appear to suffer from lack of specificity. For example, oxidation methods to remove organic phases that assume a minimal destruction of black carbons up to 400°C may underestimate the black carbon contents due to substantial oxidations taking place below this temperature. In methods where organic phases are removed by thermal treatment there is the possibility of carbonization. Finally, the assumption that black carbon is pure carbon, where, in fact, it is not, can lead to overestimates of black carbon concentrations due to its content of other components.

QUALITATIVE METHODS

Scanning Electron Microscopy

The scanning electron microscope has been the basis for the identification of the origin of small charcoals found in sediments (Scott and Collinson, 1978). Fossil woods, leaves, grasses and stems, preserved as charcoal, often show delicate three dimensional structures, such as the glandular hairs on leaves, which are characteristic of specific species. Scott and Collinson (1978) have reviewed the techniques for the disaggregation and dissolution of sediments and sedimentary rocks. HF is proposed for the dissolution of siliceous materials whereas HCl is suitable for calcareous materials. Poorly consolidated sediments can be disaggregated with hydrogen peroxide. Other disaggregation techniques previously proposed include immersion in water, boiling in sodium bicarbonate solution, and soaking in nitric acid. Any subsequent treatment, such as sieving and cleaning, should be carried out with extreme care to minimize damage to the plant structures. Scott and Collinson (1978) gold coat their mounted specimens before examination under the scanning electron microscope.

Raman Microprobe

A Raman microprobe powered by an argon/krypton ion laser identified airborne black carbon particles (Etz et al., 1979). Two bands centered around 1350 and 1600 cm^{-1} could be ascribed to the presence of elemental carbon. Further, these bands were shown to result from the laser-induced decomposition of hydrocarbons adsorbed on aerosol surfaces to carbon as well from the ambient particles taken from air. Particles of about 1 μm in diameter and larger or masses around 1 pg yielded analytical-quality Raman spectra. The identification of particulate black carbon particles can be made through Raman scattering in a modified optical microscope [Molecular Optical Laser Examiner (MOLE)] in which a laser is used as the monochromatic light source for sample irradiation (Delhaye et al., 1979). Samples of 1 nm or greater can be analyzed in this way. Two types of illumination are used, point and global (Figures A.1 and A.2).

In the former, which is known as the spectral model, the laser beam is focused through the bright field objective onto the sample and the scattered light is sent along transfer optics to a photomultiplier detector. The amplified signal is then sent to a chart recorder, TV monitor, or CRT screen. By this technique, a black carbon particle was found in the histological section of a fish liver. The identification was made by comparing its Raman spectra with those of graphite and charcoal particles (Figure A.3). In the latter, the global mode, a larger area is used, 150 to 300 μm. Here, the circular area is illuminated with a rotating laser beam that feeds the objective annular illuminator (dark field illumination system) (Figure A.2). Through selection of a characteristic radiation for

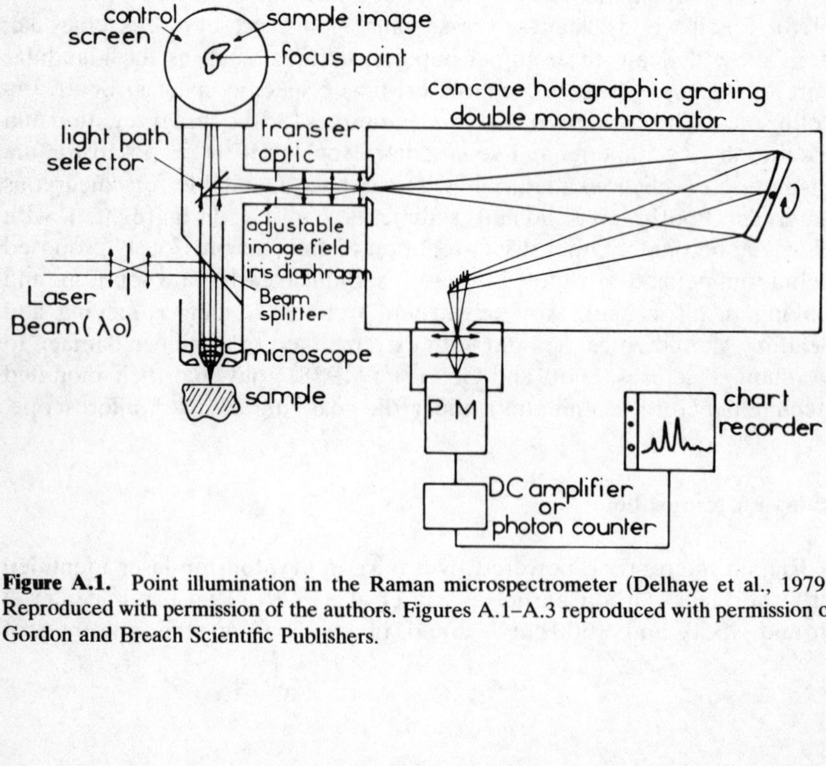

Figure A.1. Point illumination in the Raman microspectrometer (Delhaye et al., 1979). Reproduced with permission of the authors. Figures A.1–A.3 reproduced with permission of Gordon and Breach Scientific Publishers.

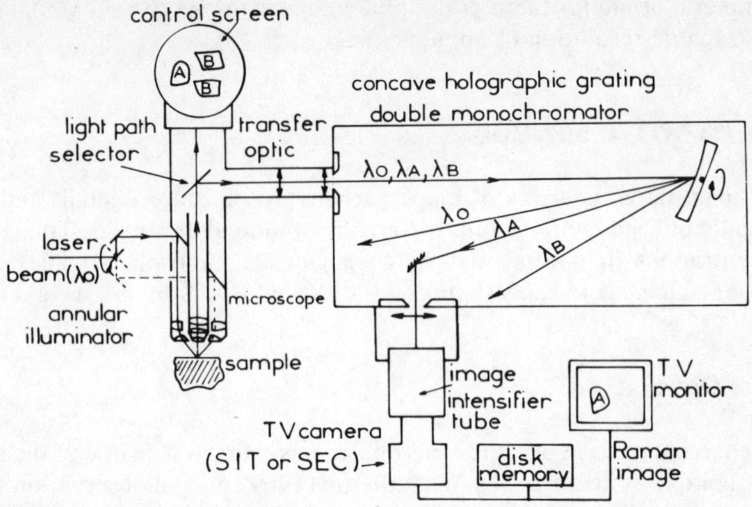

Figure A.2. Global illumination in the Raman microscope (Delhaye et al., 1979). Reproduced with permission of the authors.

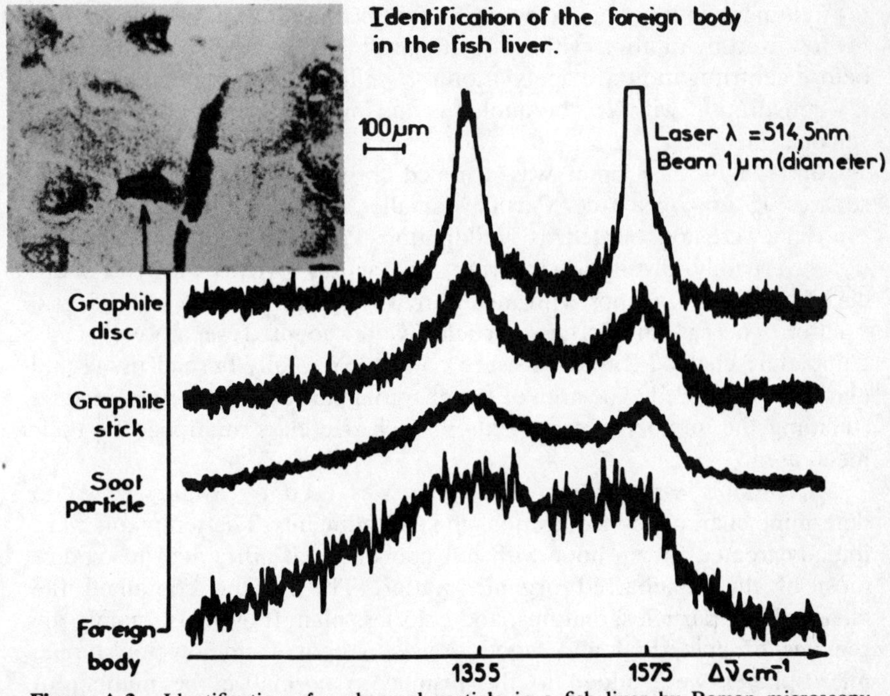

Figure A.3. Identification of a charcoal particle in a fish liver by Raman microscopy (Delhaye et al., 1979). Reproduced with permission of the authors.

the mineral of interest, a micrographic image is formed on a TV monitor with a spatial resolution of about 1 μm.

QUANTITATIVE METHODS

The quantitative analyses of black carbons in solids are difficult and all presently utilized methods suffer from one or more deficiencies. The usual concentrations in natural materials, specifically atmospheric dusts and sediments, appear to range between 0.00X and 0.X% by dry weight.

Visual Observations

The charcoal content of sediment samples was estimated, in one of the first investigations of this type, by Waddington (1969). Visual observation was used as an ancillary activity in pollen investigations in lake sediments from the Big Woods of Minnesota, USA. The pollen and charcoal were isolated from the sample by (1) initial treatment with hot 10% KOH for 2 to 5 min, (2) treatment with hot acetic acid, (3) treatment with 48% HF for 15 min, (4) followed by dilution with 95% ethanol to reduce the specific gravity before centrifugation, (5) acetylation for 1 min, (6) staining with safranin, (7) dehydration with *tert*-butanol, (8) and mounting in 200-centistokes silicone oil.

An area of 22×2 mm was scanned through a graticule grid with squares 15 μm on a side. Particles smaller than one-quarter of a grid (56 μm^2) were not counted, as Waddington (1969) claimed that they could not be readily distinguished from crystals of pyrite. Particles were classified into 11 groups with mean areas ranging from 0.5 to 27.5 grid squares whereas the larger particles were counted separately. Only completely charred fragments were counted; partially burned tissue and chitin were ignored. The area of the charcoal particles was determined by summing the number of fragments in each size class multiplied by their mean area.

Microscopic examination of isolates was used by Swain (1973) to determine charcoal concentrations in lake sediments. The sediments were initially treated for an hour with hot concentrated nitric acid to oxidize most of the noncharred organic matter. The residue contained the charcoal, silt particles, diatoms, and colorless plant fragments, mainly the remains of epidermal and wood tissues. About 1 million-polystyrene microspheres were added to the residue to normalize the number of charcoal particles counted. Each of the initial sediment samples used by

Swain (1973) represented 10 yr of deposition. The chronology was ascertained by varve counting. The preparation of residue and microspheres was washed repeatedly and finally mounted in glycerine on a slide with a 22 × 22-mm cover glass. It was claimed that charcoals could be distinguished from unburned plant materials by color—charcoals varied from brown to black, whereas the uncharred materials remained colorless after the nitric acid treatment.

The surface area of each charcoal fragment was estimated by use of a graticule grid with squares 13 μm on each side. Fragments smaller than one half of a grid square were ignored whereas particles >35 squares were measured individually. Particles were placed into seven size classes: 0.5–4.5; 4.5–9.5; 9.5–14.5; 14.5–19.5; 19.5–24.5; 24.5–29.5; and 29.5–34.5 squares, depending upon the number of squares covered by the particle. Twenty equally spaced traverses were made across each slide. The total area of charcoal recorded for all size classes was computed by summing the product of the number of pieces and the midpoint of the class.

The results are reported as surface area per square centimeter per year (area/cm^2/yr). The number of years represented on each slide is determined by the ratio of the number of microspheres counted per slide divided by the total number of microspheres added to the 10-yr sample. The surface area involved comes from that of the coring device. No mention is made by the author of the precision of the analyses.

Cwynar (1978), using the counting methods of Swain (1973), measured the precision of his technique by determining the charcoal content of a lacustrine sediment. Three replicate samples were used from a stratum that was deposited between 1260 and 1269 A.D.

Run	Charcoal Fluxes in 10^6 $\mu m^2/cm^2/yr$		
	A	*B*	*C*
1	3.23	4.49	3.37
2	3.31	2.89	3.24
3	3.41	5.01	3.07
4	2.94	4.01	3.33
5	2.65	3.13	3.99
Mean	3.11	3.90	3.40
Standard deviation	0.31	0.90	0.35

Byrne et al. (1977) used a more complex isolation procedure to recover charcoal from sediments. Samples were subjected to the following treatments: HCl (10%), 5 min; KOH (10%), 5 min, 8 to 10 washes with distilled water and a short centrifugation (1 min at 2000 rpm) to remove the colloidal material; HF (concentrated) 24 hr; HNO_3 (concentrated) 4 min, and acetolysis. The residue was mounted in silicon oil and the areas of charcoal fragments were determined with the aid of an eyepiece graticule containing 25×25-μm squares. Only charcoals with obvious cellular structures were counted.

Oxidation to Carbon Dioxide

One of the first techniques to quantify black carbon isolates by oxidation to carbon dioxide was carried out by Mueller et al. (1971). These workers recognized that the atmospheric carbon contents would be in the forms of elemental carbon, biological particles, or carbonate. They analyzed for two fractions: carbonate and noncarbonate carbon. The former was determined by an acidification of aerosols collected on glass fiber filters or aluminum foils and, subsequently, assaying the evolved carbon dioxide. The noncarbonate carbon, which contained the black carbon, was determined by heating the residue to 900°. Later work has sought to define more precisely the composition of the noncarbonate fractions.

As an example, total carbon and organic carbon plus carbonate carbon were determined on atmospheric particulates through preferential oxidation (Heisler et al., 1980). The samples were collected on quartz fiber filters. The materials were oxidized in a platinum boat with manganese dioxide. All carbonaceous materials were presumed oxidized at 850°, whereas the organic phases and the carbonate carbonates were destroyed at 550°C. Elemental carbon is then given by difference of the total carbon and the organic plus carbonate carbon. The particulates contained negligible carbonate. The generated carbon dioxide was hydrogenated catalytically to methane and measured in a gas chromatograph using a flame ionization detector. The filters used (Pallflex QAST quartz fibers) contained no detectable elemental carbon and 5.6 γ/cm^2 of organic carbon.

A slight modification of this technique was proposed by Johnson and Huntzicker (1979) who acidified the filter initially to remove the carbonates. Their lower limit of detection was <1 μg of black carbon on the filter.

A modification of this technique has been developed at the General Motors Research Laboratory (Muhlbaier and Williams, 1982). Atmospheric black carbons collected on glass or quartz fiber filters were injected into a furnace, flushed with helium, at 650°C to volatilize off the organic carbon components which were then catalytically oxidized to carbon dioxide. Air was then introduced to the system and the remaining black carbon was then combusted to carbon dioxide. One problem with this technique is that some of the organic matter may be converted to black carbon and hence its concentration can be overestimated. In order to reduce charring, the technique can be altered whereby the organic phases are removed by heating to 350°C in air. The total carbon can be determined by the original method.

The problem of charring through volatilization of organics in helium atmospheres was also addressed by Ogren et al. (1983) who utilized a low temperature oxidation of the organic matter with alkaline peroxide, a technique based on that of Smith et al. (1975). Their samples were rainwaters, high in organic carbon. Following a filtration to remove particles larger than 5 μm (primarily organic phases with small contributions of black carbon), the filtrate was treated with basic hydrogen peroxide. Not all of the organic matter was destroyed so a subsequent oxidation at 900°C in a helium atmosphere was used. The evolved carbon dioxide from a combustion at 900°C was measured with a nondispersive infrared analyzer.

The measurement of the organic matter pyrolyzed to black carbon in the thermal analyses has been made by Huntzicker et al. (1982). A correction can then be applied to the thermal determination of black carbon in aerosols containing organic phases. A two step oxidation of the organic carbon, which has been captured on glass or quartz fiber filters, involves oxidation at 350°C in O_2 and at 600°C in He. The volatilized organic carbon is oxidized to CO_2 and reduced to CH_4, which is measured in a flame ionization detector of a gas chromatograph. The reflectance of the filter is continuously measured by a He–Ne laser system and decreases during the combustion of the organic phases through their pyrolytic conversion to elemental carbon. The correction is ascertained by determining the amount of black carbon necessary to return the filter to its original reflectance. In aerosols of U.S. cities, the pyrolytically converted carbon could achieve values of over half of the ambient black carbon.

The thermal techniques, pyrolysis in He, and oxidation in air gave systematically different results, depending upon the amounts of carbonization and oxidation of the black carbon (Table A.1). The conclusion of Cadle and Groblicki (1982) is that the methods they tested give highly correlated results. For use on diesel particulates, where the organic components are low, there will be minimal errors. For charcoals, on the other hand, the techniques can introduce high errors due to carbonization. They recommend step-wise thermal analysis, incorporating an oxidation

Table A.1. Comparison of Measurements of Black Carbon Contents in Denver Atmospheric Particulates by a Variety of Techniques (Cadle and Groblicki, 1982)[a,b]

Method	Regression Line[c]
350°C oxidation followed by total carbon analysis	$y = 0.81 C_{abc} + 5.6$
350°C oxidation followed by 650°C pyrolysis–oxidation	$y = 0.71 C_{abc} - 1.14$
Extraction followed by total carbon analysis	$y = 0.82 C_{abc} + 7.7$
Extraction followed by 650°C pyrolysis–oxidation	$y = 0.71 C_{abc} - 2.4$
Nitric acid digestion followed by total carbon analysis	$y = 0.81 C_{abc} - 0.71$
Nitric acid digestion followed by 650°C pyrolysis–oxidation	$y = 0.57 C_{abc} - 3.0$

[a] Reproduced with permission of the authors and Plenum Press.
[b] The apparent black carbon content (C_{abc}) is related to the measured value (y). The former was determined by pyrolysis–oxidation at 650°C.
[c] y and C_{abc} are in units of (μg/cm^2).

procedure to minimize carbonization. Optimal separation of organic compounds from black carbon involves a 350°C air oxidation followed by pyrolysis in He at 950°C (Cadle et al., 1983). These investigators have developed an automated carbon analyzer that can allow 23 samples to be analyzed during 8 hr of unattended operation (Cadle et al., 1980).

A second problem with the oxidation technique arises from the possibility of the oxidation of black carbon during the removal of the organic phases. This uncertainty has been addressed by Tanner et al. (1982). These investigators used a two-step technique, the first of which is a rapid heating at 400°C in a helium carrier gas. Then, the black carbon is oxidized at 700°C in a 10% O_2/He mixture. The evolved carbon is converted to carbon dioxide on a copper oxide catalyst, purified and analyzed by nondispersive infrared spectrometry. Very rapid heating of the samples was employed, about 600°C/min, to minimize the effects of carbonization. The investigators argue that at elevated temperatures flash volatilization or desorption of organic phases from the sample overrides dehydrogenation and dehydration reactions that lead to carbonization. On the basis of their results, the possibility still exists of a conversion of up to 10% organic matter to carbon. Since the ratio organic carbon/total carbon increased linearly with pyrolysis temperature, there may be loss of some of the black carbon during the removal of the organic phases. Clearly, the distinction between organic and elemental carbon by this method is arbitrary. The investigators chose the temperature of conversion at 400°C because at this temperature the slope of the ratio *versus* pyrolysis temperature is close to zero.

Black carbon has been determined in atmospheric aerosols after collection on a high volume glass fiber filter and subsequent oxidation (Kukreja and Bove, 1976). The glass filter mat is destroyed with HF, as well as the silicate minerals in the dust. The organic compounds are removed with ammonium hydroxide, nitric, and hydrochloric acid treatments.

Twenty-four-hour periods are used for the collection of air upon an 8 × 10-in. glass filter. Half of the filter can be used for the spiking with a standard (lampblack) or both halves of the filters can be used to determine the precision of the technique. The filter half is cut into 20 pieces which are added one by one to 15 mL of HF in a 100-mL platinum dish. The reaction is rapid. The mixture is dissolved in 15 mL of distilled water and transferred to a graduated 500-mL Pyrex beaker. Sixty milliliters of concentrated ammonium hydroxide are carefully added and followed by 70 ML of concentrated nitric acid, along with a few glass beads. After evaporation to about 100 mL, 30 mL of nitric acid are added and the mixture again evaporated to 100 mL. The mixture was then diluted with 250 mL of distilled water and 10 mL of HCl. The resulting mixture was heated to boiling and filtered through a tared, prefired (at 1000°C) Gooch crucible. The contents of the crucible were washed with 300 mL of a warm 1% nitric acid solution.

The crucible and its contents were dried at 150°C to constant weight, which took about 2 hr. The crucible was weighed, heated to 700°C for 2 hr, cooled, and reweighted. The difference in the two weights represents the amount of free carbon that has been oxidized to carbon dioxide. There is about 1 mg of free carbon in the filter paper halves. Control samples with 21 to 82 mg of lampblack added were analyzed with a recovery of carbon of at least 1.4 mg or better.

Some nine samples were collected between March and November 1974 from the roof of the Copper Union (20 m) in New York City. The total weight of the particulate matter collected during a 24-hr period varied between 111 and 213 mg. The black carbon varied between 5 and 16%. Inasmuch as only nine samples were collected, Kukreja and Bove (1976) offered no interpretation of the data.

Infrared Absorption

Black carbons, when ground in air for periods of parts of a day to days, produce a characteristic infrared spectrum (Smith et al., 1975). In the range of 2500–1000 cm^{-1}, absorption bands appear at 1720, 1580, and 1240 cm^{-1} (Figure A.4). The 1580-cm^{-1} absorption band is preferred for analytical work inasmuch as it is the most characteristic and the most intense.

Surface oxidation of the black carbon during grinding by oxygen appears responsible for the development of the bands. Black carbon ground in air develops the characteristic spectrum whereas when ground in nitrogen it does not (Figure A.5). The specific absorption of the black carbons appears to be constant after a 24-hr grinding period.

The principal interference in the technique is due to the introduction of water which has a 1625-cm^{-1} absorption band (Smith et al., 1975). Water can be introduced during the mixing of the carbon isolate with KBr used to prepare the pellet for infrared assay. To minimize this interference, samples are prepared using thoroughly dried KBr, mixed by grinding for 30 s in a stainless steel vial, and dried overnight at 120°C prior to pelleting.

The isolation of the black carbon from sediments or atmospheric dust prior to infrared absorption assay involves the dissolution of siliceous

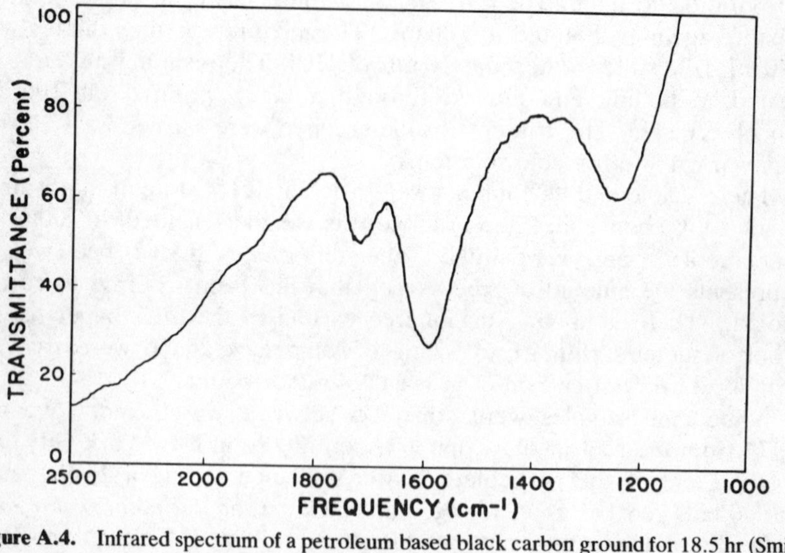

Figure A.4. Infrared spectrum of a petroleum based black carbon ground for 18.5 hr (Smith et al., 1975). Reproduced with permission of the authors and the American Chemical Society.

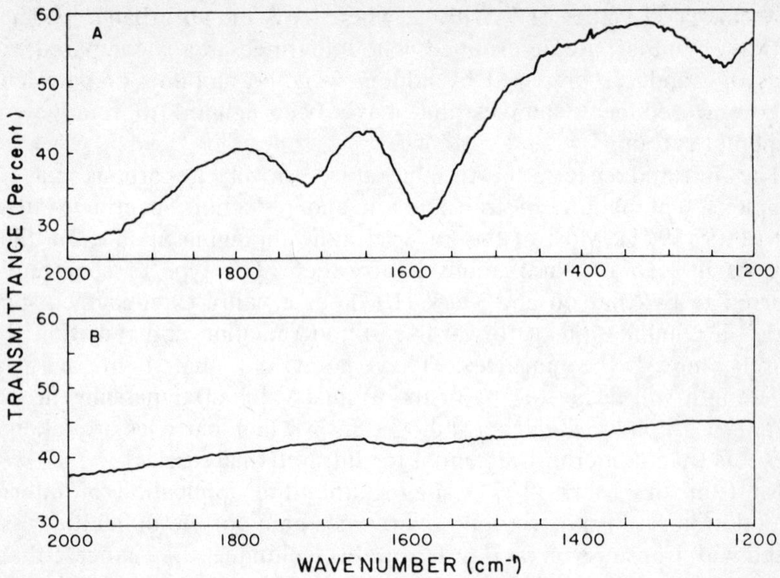

Figure A.5. Comparison of a petroleum based black carbon ground in air (A) and in nitrogen (B) (Smith et al., 1975). Reproduced with permission of the authors and the American Chemical Society.

materials, carbonates, and organic phases. Acid soluble materials are removed by treating 1–10 g of sample with hot 6 *N* HCl for 2 hr. The soluble salts are then removed by repeated washings and centrifugation with distilled water. The silicates in the resulting residue are put into solution through digestion with 28% HF for about 10 days. The remaining solids are then washed with hot 6 *N* HCl to dissolve any fluorides formed during the previous treatment. Finally, this isolate is washed with distilled water and dried prior to pelleting.

For samples with organic phases, the initial treatment begins with their destruction. This is accomplished by adding 150 ml of 6 *N* KOH cautiously to 10 g of dried sediment or dust (Griffin and Goldberg, 1975). After dispersion of this mixture by ultrasonication, 20 mL of 30% H_2O_2 are added in increments of 2 mL, so that foaming is held to a minimum. By such a treatment, the organic phases are removed to a point where they do not interface with the infrared analysis. The residue is washed with distilled water and the acid dissolution steps are then taken. The residue after these treatments includes the black carbons and such resistant minerals as pyrite, anatase, rutile, and zircon.

Pelleting of the KBr sample mixture is carried out under a pressure of 700 kg/cm² for 10 min *in vacuo*. The absorption band is measured at

1580 cm^{-1} and values of $\int A(v)\,dv$, where A is the absorbance and v is the wave number, are determined. The integrated area is compared with areas of standards prepared by adding weighed amounts of petroleum carbon to sediment samples that have been ignited to remove any elemental carbon.

The infrared spectra in the literature for black carbons may be complex combinations of transmission and reflection spectra (Mattson and Mark, 1971). Most of the infrared light impinging upon even dilute samples in KBr or Nujol mulls is absorbed. The type of spectrum is referred to by Mattson and Mark (1971) as a "diffuse reflectance spectrum," a combination of forward-scattered radiation and radiation that entirely misses the particles. They point out that light of 5-nm wavelength will decay to 1% of its original value after passing through 3.7 nm of graphite. Thus, a grinding step in which particles are taken to sizes less than a micron is essential for infrared analysis.

Mattson and Mark (1971) suggest that the application of internal reflection infrared spectroscopy overcomes many of the difficulties associated with transmission modes. Using this technique, they indicate that it is experimentally possible to obtain high contrast spectra with great resolution. Clearly, the application of this technique will be most important in the short wavelength regions.

Absorption in the Visible

Visible light absorption by black carbons provides the bases of a technique for assays of urban atmospheric dusts (Rosen et al., 1980). Samples are collected on cellulose membrane filters (Millipore Type RATF, 1.2-μm nominal pore size, 47 mm in diameter) which are used for the optical attenuation measurements. The optical attenuation, A, is defined by

$$A = -100 \ln I/I_0$$

where I_0 is the intensity of the light transmitted through a blank filter disc and I is that transmitted through the sample at a wavelength of 0.63 μm.

On the basis of arguments posed by Rosen et al. (1979) the visible absorption is due to the black carbon component of the aerosols. First of all, the optical attenuation has a $1/\lambda$ wavelength dependence to within 20% over the visible spectrum (0.45–0.70 μm). Such a situation is consistent with the particles having a constant, imaginary index of refraction. Rosen et al. (1979) indicate that soots produced from acetylene and propane burning possess essentially constant indices of refraction throughout the visible region. Further, they point out that of the 13,000 organics listed in the *54th Edition of the Chemical Rubber Handbook*, only 5 are gray or black in appearance, a characteristic that would provide a similar behavior to the aerosols.

In addition, the temperature stability of the aerosols corresponds to those of poorly crystalline graphite. The particles after heating for 30 min in air, are stable to a temperature of approximately 400°C. At this temperature, oxidation apparently begins. The particles essentially disappear by 500°C. This experiment eliminates metal oxides as important components.

Finally, the particles show no solubility in nonpolar or polar organic phases. Soxhlet extraction into benzene and methanol–chloroform indicated no removal of the atmospheric dusts, a situation to be expected for black carbons but not for most organics.

An intercomparison between the integrating plate and the laser transmission methods for determination of aerosol absorption coefficients has been carried out and a high correlation between the carbon content of the aerosol in urban regions has been determined (Sadler et al., 1981). The integrating plate method is based upon the transparency of a Nucleopore filter with the aerosol compared to that of a clean filter to light of 525-nm wavelength. The laser transmission method is an extension of this technique to other filter types (Millipore, quartz fiber, teflon, etc.) where the transmission of a 633-nm He–Ne laser beam is used. The relative transmission through a filter loaded with aerosol and through a clean filter is determined.

The transmission data was compared with the total black carbon contents of the filters determined by the CO_2 evolution method (Mueller et al., 1971). The blanks of the filters ran about $0.5\gamma\ C/cm^2$. Sampling sites ranged from congested highway tunnels to a remote location on a western foothill. Absorption coefficients varied over three orders of magnitude and the total carbon concentrations were from 5 to 90 γ/m^3. There is a 0.95 correlation coefficient between the integrating plate and laser transmission methods and a 0.96 correlation coefficient between the concentration of black carbon and the absorption on Nucleopore filters. Clearly, the two methods are comparable and correlated. But perhaps more important is the correlation between the absorption coefficients and the carbon loading, a correlation that has been observed in other work.

The intensity of the Raman band near $1600\ cm^{-1}$, attributed to black carbon, and the visible absorption covaried (Figure A.6) for urban air particulates collected in California cities. A reasonable explanation of this result is that the absorptivity is due to black carbons. Rosen et al. (1979) point out that the chemical composition of these samples were

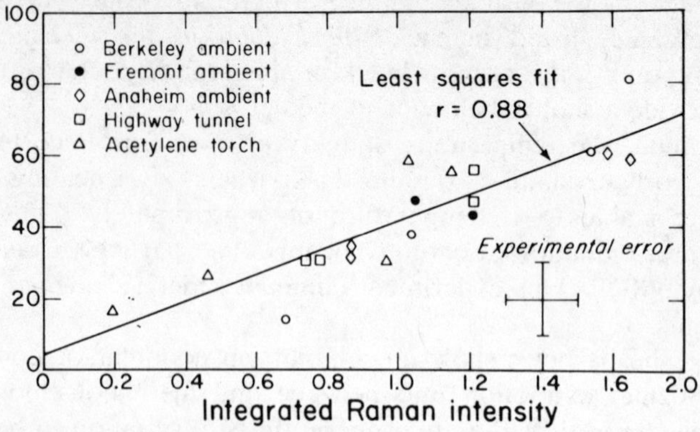

Figure A.6. Integrated Raman intensity of the 1600-cm^{-1} band versus the percentage optical attenuation at 6323 Å for ambient black carbon and acetylene soots (Rosen et al., 1979). Reproduced with permission of the authors.

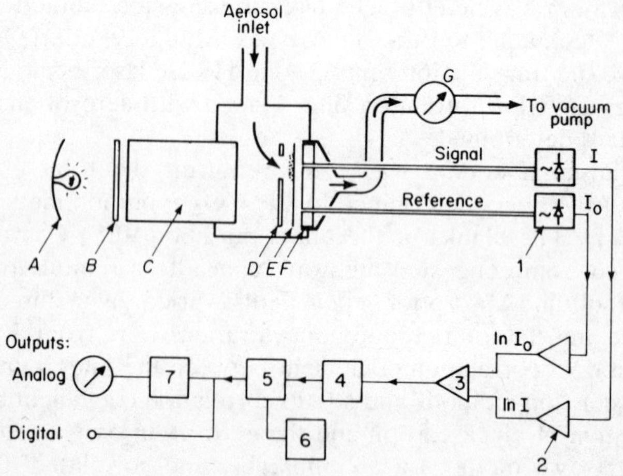

Figure A.7. Block diagram of the aethalometer: *A*, light source; *B*, 530-nm bandpass filter; *C*, quartz light guide: *D*, transparent mask; *E*, filter with particles collected on portion underneath hole is mask; *F*, filter support with optical fibers set in *G*, flowmeter; (1) silicon photodetectors; (2) logarithmic amplifiers; (3) difference amplifier giving output proportional to ln (I/I_0); (4) A/D converter; (5) storage and subtraction; (6) variable time base; (7) D/A converter (Hansen et al., 1984). Reproduced with permission of the authors and Elsevier Scientific Publishers.

quite different. For example, the Pb and Fe concentrations varied over two orders of magnitude.

Also, Rosen et al. (1979) indicate that the *total* carbon concentration in aerosols from California cities strongly correlates with the black carbon component as measured by the visible absorption. This suggests that the black carbons are a major fraction of the aerosols. The aerosols are apparently stable to ozone oxidation in the atmosphere inasmuch as there is no measurable change in the black carbon to total carbon ratio during peak ozone periods.

An optical attenuation technique proposed by Rosen et al. (1978), Rosen et al. (1982), and Hansen et al. (1979) compares the transmission of a 633 nm He/Ne laser beam through a filter with atmospheric particulates upon it to that of a blank filter. Filters used include Millipore cellulose membranes and quartz fiber substrates.

The percentage attenuation shows a strong covariance with the integrated Raman intensity (Figure A.6) for a variety of samples for California, including acetylene soots and airs from highway tunnels. Thus, in principle, both methods can be used to give measures of black carbons in the atmosphere.

The A and the concentration of black carbon (BC) covary according to the equation

$$BC = A/s$$

where s is the specific attenuation for black carbon (Gundel et al., 1984). s has a value of $25.4 \pm 1.7\ cm^2/\mu g$ for black carbon emissions from such sources as diesel vehicles, propane soot, and natural gas soot as well as for atmospheric aerosols. The particulates studied had the properties of black carbons—high temperature stability with only minimal oxidation up to 400°C and insolubility in a wide variety of solvents.

Continuous measurements of black carbons in the air can be made with an instrument, called an "aethalometer," which measures the attenuation of a light beam (stabilized incandescent 530-nm lamp) transmitted through a filter upon which particle-laden air is drawn (Hansen et al., 1984). The rate of deposition of the black carbon on the filter is related to its concentration in the air (Figure A.7). The technique assumes that the absorption of light is due to the black carbon in the aerosol.

Calibration is carried out by a quantitative determination of the black carbon at the end run and an integration of the output of the instrument. Through these two measurements a calibration factor is obtained.

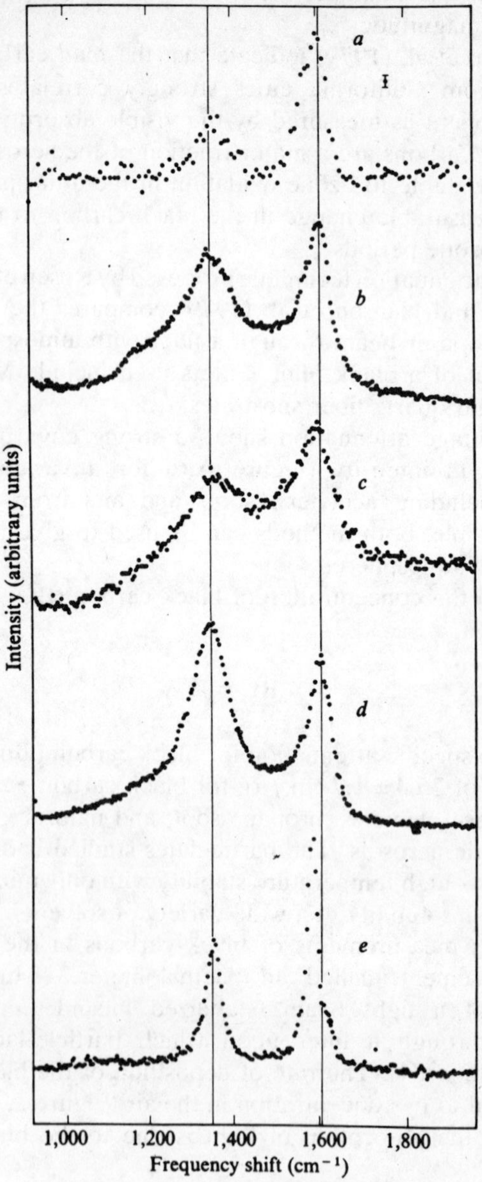

Figure A.8. Raman spectra of (*a*) ambient air, (*b*) automobile exhaust, (*c*) diesel exhaust, (*d*) activated carbon, and (*e*) polycrystalline graphite (Rosen and Novakov, 1977). Reproduced with permission of the authors and *Nature*.

Raman Scattering

Raman scattering has been used to characterize soot issuing from diesel and gasoline exhausts and from ambient airs (Rosen and Novakov, 1977). Samples were irradiated with a beam from an argon ion laser, operating at 1 W at a wavelength of 5145 Å. The Raman output was recorded by a photomultiplier tube cooled to −20°C and the emitted photon pulses were counted on a multichannel analyzer. To minimize heating effects, the samples were rotated at 1800 rpm.

The initial studies indicated that the engine exhausts and ambient airs from St. Louis, Missouri, USA, contained particles with physical structures similar to activated carbon (Figure A.8). Further, they suggested that these carbon species may be a major component of both the exhausts and airs. The investigators surveyed literature data to yield crystallite sizes from the observed Raman modes: ~50 Å for the exhausts and ~100 Å for the air.

To be able to quantify carbon concentrations in the aerosol isolates, measurements of Raman cross sections, optical absorption cross sections, and particle size effects must be ascertained.

Photoacoustic Spectroscopy

Photoacoustic spectroscopy was employed by Yasa et al. (1979) to assess the results of Rosen et al. (1978) who used optical absorption techniques to establish that the primary constituent of atmospheric aerosols from urban areas is black carbon. There was some uncertainty as to whether the optical absorption technique measures exclusively absorption rather than the scattering of the aerosol particles. Theoretically, if the output of both techniques covary, then it follows that the optical attenuation measurements are controlled by the light-absorption properties of the aerosols. This is a consequence of the fact that the the photoacoustic signal reflects the heat generated by absorption.

The apparatus, shown in Figure A.9, utilizes a He–Ne laser that is focused upon the aerosol sample collected on a Millipore filter and mounted on pyrex. The laser operates with 0.5 mW of power and with a modulation frequency of 20 Hz. The assumption is made that the optically absorbing component of the aerosol is black carbon.

Optical attenuation and photoacoustic measurements were made on a collection of urban particulates from California, Colorado, New York, and particles collected in a highway tunnel and from an acetylene torch

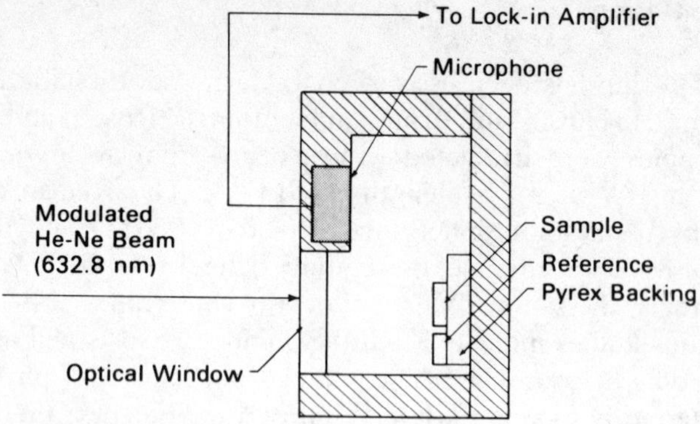

Figure A.9. The experimental arrangement for the photoacoustic determination of black carbons (Yasa et al., 1979). Reproduced with permission of the authors and *Applied Optics*.

(Figure A.10). The least-squares fit of the data provides a correlation coefficient of 0.98 and a slope of 1.03. These results strongly indicate that the absorbing component of urban aerosols is primarily graphitic carbon.

A comparison of analyses of butane soots on glass fiber filters by photoacoustic spectroscopy and reflectance indicate both techniques are

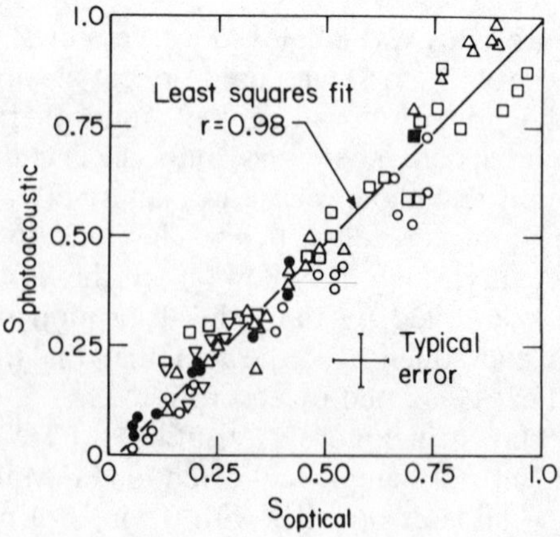

Figure A.10. Photoacoustic signals and optical signals on urban particulates from Fremont, California ▽; Anaheim, California □; Denver ○; New York City △; highway tunnel ■; and an acetylene torch ● (Yasa et al., 1979). Reproduced with permission of the authors and *Applied Optics*.

competitive (Delumyea and Mitchell, 1983). The black carbon loadings were determined by conventional combustion techniques.

Both techniques displayed a linear response from 0 to 15 μg of carbon per square centimeter. The reflectance signal reaches a plateau at about 17 $\mu g/cm^2$ at which level the surface is completely covered with black carbon. On the other hand, the photoacoustic signal continues to increase after this amount of loading although at a reduced slope.

Deuteron Activation Analysis

A nondestructive technique for atmospheric aerosols utilizing charged particle activation of the sample has been described by Clemenson et al. (1980). The reaction $^{12}C(d, n)^{13}N$ produces a radionuclide (^{13}N) with a 10.0-min half-life, a positron emitter that is detected by its 0.511-MeV annihilation gamma rays. The University of California Lawrence Laboratory 88-in. cyclotron facility was used for the irradiations.

The energy of the incident radiation was determined by a survey of the potentially interfering reactions:

Reaction	Threshold Energy (MeV)
$^{12}C(d, n)^{13}N$	0.33
$^{14}N(d, t)^{13}N$	4.91
$^{14}N(d, dn)^{13}N$	12.06
$^{16}O(d, \alpha n)^{13}N$	8.37

The latter two reactions will be eliminated if the energy of the incident deuterons is <8 MeV. The (D, t) reaction on ^{14}N has a cross section of 1.3 mb at 7.6 MeV. This is substantially less than that of the reaction involving ^{12}C with a value of 70 mb at the same energy. In addition, atmospheric aerosols usually contain 1–10 times more carbon than oxygen. Thus, an irradiation energy of 7.6 MeV was utilized in the analyses.

The aerosols were collected on a silver-membrane filter that was directly irradiated with deuterons. The silver is only slightly activated due to its Coulomb barrier for the deuterons. Stacked samples were bombarded. The stack contained a polystyrene foil, used as a carbon standard, a filter sample, and aluminum foils. The aluminum foils were used to degrade the deuteron energy to the desired value. Typical loadings of the silver filter were 250 mg/cm^2. The stack was irradiated for 2 min with a beam intensity of 0.5 μA.

The decay of the annihilation gamma rays was followed in a counting apparatus that included a 60 cm^3 Ge(Li) detector with a resolution of 2.0 keV at the 1.33-MeV gamma ray of ^{60}Co and a 4096 channel computer controlled analyzer.

The decay of the positrons was followed for 2 to 3 hr. The aerosol samples had gamma rays in the region of 0 to 1 MeV from the activation of silver (^{107}Ag, ^{108}Ag, and ^{107}Cd) as well as of aluminum (^{27}Mg). The decay curve could be resolved into four components: (1) a 10.0-min component from the decay of the ^{13}N produced from the ^{12}C; (2) a low 24.1-min component due to the $^{107}Ag(d, t)$ 106 reaction; (3) a 109.8-min component due to ^{18}F produced from the $^{17}O(d, n)$ ^{18}F and $^{18}O(d, 2n)$ ^{18}F reactions; and (4) a 6.5-hr component due to ^{107}Cd produced from $^{107}Ag(d, 2n)$ ^{107}Cd reaction in the silver filter. By comparison of the computed net ^{13}N activity with that of the polystyrene standard, the net carbon content of the filter could be obtained.

The results were compared with those from the combustion of the samples. The ratio, carbon by deuteron activation to carbon by combustion, had a value of 1.01 ± 0.10 over a range of carbon contents of 0.6 to 268 μg of carbon per square centimeter, of filter.

Selective Extraction

The separation of the black carbon from the organic phases in environmental samples has been most vexing (Cadle and Groblicki, 1982). Some of the high molecular weight organic compounds may have compositions that blend into that of black carbon. The separation techniques are empirical and the resultant measurement of black carbon will be technique dependent. Cadle and Groblicki (op. cit.) have compared a variety of extraction techniques, some of which had been previously proposed (Table A.2). Both a 350°C oxidation and a 2-hr nitric acid extraction removed 67–68% of the carbon, presumed to be organic carbon, from atmospheric particulates collected in Michigan while extraction techniques removed a maximum of 54% of the carbon. There appears to be about 14% of the organic matter that is refractory to extractive techniques. Appel and his associates (Appel et al., 1976, 1979) have selectively removed from atmospheric aerosols various carbon containing

Table A.2. Removal of Organic Carbon from Six Atmospheric Particulate Samples (Cadle and Groblicki, 1982)[a]

Method	Average Percentage Carbon Removed
Extraction	
Fluorinert FC-78	20
Water	27
Hexane	32
o-Dichlorobenzene (O_2 atmosphere)	40
Dichloromethane	40
Toluene	47
Benzene followed by 1:2 methanol–chloroform	51
4:1 Benzene–ethanol	54
Digestion	
Nitric acid–2 hr on steam bath	68
Nitric acid[b]–24 hr–batch 1	73
Nitric acid[b]–24 hr–batch 2	85
Thermal analysis	
180°C vacuum	31
650°C pyrolysis	58
350°C vacuum	62
350°C oxidation	67

[a]Reproduced with permission of the authors and Plenum Press.
[b]Leave 0.5 hr on hot plate, dilute, and let stand overnight. Batch 1 and 2 are repeats from the same filters.

compounds to yield residues which they submit are measures of elemental carbon. The particulates were treated both with polar solvents (1:2 V/V methanol–chloroform) and nonpolar solvents (cyclohexane and benzene) to yield a residue ascribable to black carbon. Inorganic carbonates were taken into account, however, carbon in such polymers as rubber and viable particles were not. Thus, their results are upper limit values. The residues were converted to carbon dioxide that was measured by gas chromatography. The carbon measured as the elemental form was the most abundant species in the aerosols collected at some California sites.

Twenty-hour extractions of diesel exhaust particulates with 1:1 (V/V) toluene/1-propanol yielded an amount of residue that in most cases compared favorably with the total black carbon content measured by thermal optical techniques (Japar et al., 1984). Any discrepancies could be attributed to compounds of Fe, S, Al, Si, and Ca in the exhaust particulates.

Some digestion techniques do dissolve black carbon (Cadle et al., 1983). Substantial amounts of black carbon in atmospheric particulates can be removed by potassium persulfate and by strongly oxidizing, acidic solutions. These authors argue that digestion methods offer no advantages over thermal methods for the separation of black carbons from organic materials.

Reflectance

The reflectance of light from filter surfaces has been used to measure the amount of black carbon collected from known volumes of air (Delumyea et al., 1980; Macias and Chu, 1982). Instruments for continuous assay in the field and for laboratory analyses have been developed. Both irradiate with tungsten filament lamps at 45° and pick up the reflected radiation at right angles with a photodetector (Figure A-11).

The method was calibrated with the black carbon assays made by the measurements of gamma rays emitted through proton bombardment, the so-called GRALE (gamma ray analysis of light elements) technique (Macias et al., 1979, Macias et al., 1978, and Macias and Chu, 1982). This technique normally has been used to measure total carbon (organic carbon, black carbon, and carbonate). However, by heating the samples of 300°C, the organic carbon is allegedly destroyed and, in the absence of carbonate, black carbon is uniquely determined on the residue. The

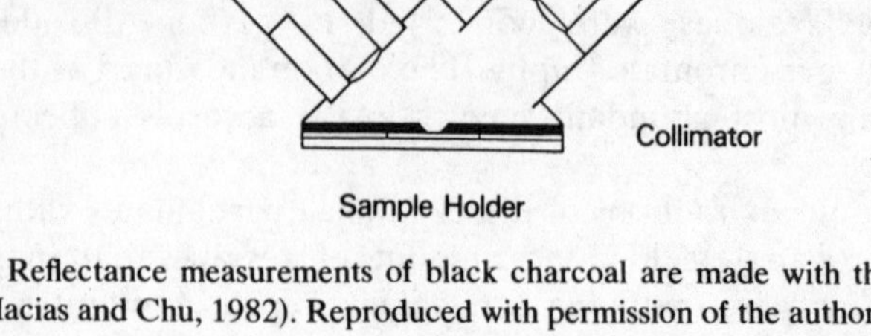

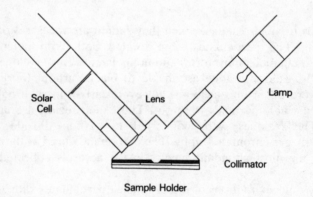

Figure A.11. Reflectance measurements of black charcoal are made with the above type instrument (Macias and Chu, 1982). Reproduced with permission of the authors and Plenum Press.

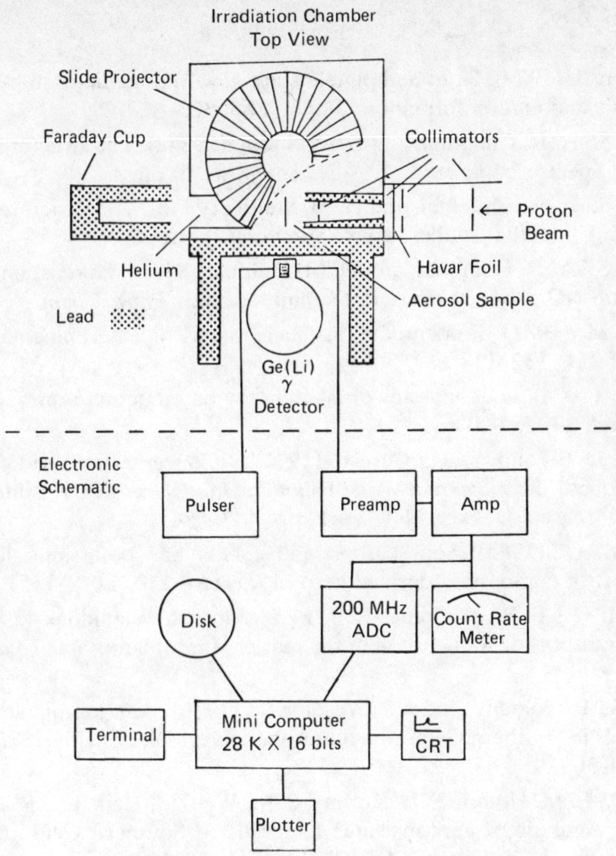

Figure A.12. Schematic diagram of the sample irradiation chamber and the electronics used for black carbon analyses by the GRALE technique (Macias et al., 1978). Reproduced with permission of the authors and the American Chemical Society.

calibrations were carried out with soots produced from the combustion of butane. The solids were heated to 300° to remove any organic phases.

In the GRALE technique, the aerosol samples, collected upon SiO_2-type filters, are bombarded for 1000 s with 7.0-MeV protons (Figure A.12). The first excited states of carbon, as well as those of nitrogen oxygen, and sulfur, are produced. The emitted gamma rays are measured on a Ge(Li) crystal detector. There are three gamma rays of carbon that are diagnostic of the element and are used in the analytical scheme.

REFERENCES

Abrahamson, J. (1977). Saturated platelets are new intermediates in hydrocarbon pyrolysis and carbon formation. *Nature* **266**, 323–327.

Akhter, M. S., A. R. Chughtai, and D. M. Smith, (1984a). The structure of hexane soot. I. Spectroscopic studies. *Appl. Spectrosc*. In press.

Akhter, M. S., A. R. Chughtai, and D. M. Smith (1984b). The structure of hexane soot. II. Extraction studies. *Appl. Spectrosc*. In press.

Akhter, M. S., A. R. Chughtai, and D. M. Smith (1984c). The reaction of hexane soot with NO_2/N_2O_4. Manuscript submitted to *J. Phys. Chem*.

Alexander, M. (1981). Biodegradation of chemicals of environmental concern. *Science* **211**, 132–138.

Alvin, K. L. (1974). Leaf anatomy of Weischselia based upon fusainized material. *Paleontol*. **17**, 587–598.

Andelman, J. B. and S. C. Caruso (1971). "Concentration and Separation Techniques." In *Water and Water Pollution Handbook*, Vol. 2. Edited by L. L. Ciaccio. Marcel Dekker, New York, pp. 483–584.

Andreae, M. O. (1983). Soot carbon and excess fine potassium: long range transport of combustion derived aerosol. *Science* **220**, 1148–1151.

Andreae, M. O., T. W. Andreae, R. J. Ferek, and H. Raemdonk (1984). Long range transport of soot carbon in the marine atmosphere. *Sci. Total Environ*. **36**, 73–80.

Appel, B. R., P. Colodny, and J. J. Wesolowski (1976). Analysis of carbonaceous materials in Southern California atmospheric aerosols. *Environ. Sci. Technol*. **10**, 359–363.

Appel, B. R., E. M. Hoffer, E. L. Kothny, S. M. Wall, M. Haik, and R. L. Knights (1979). Analysis of carbonaceous materials in Southern California atmospheric aerosols. 2. *Environ. Sci. Technol*. **13**, 98–104.

Baker, R. T., M. A. Barber, P. S. Harris, F. S. Fentes, and R. J. Wirte (1972). Nucleation and growth of carbon deposits from the nickel catalyzed decomposition of acetylene. *J. Catal*. **26**, 51–62.

Baldwin, A. C. (1982). "Reaction of gases on prototype aerosol particle surfaces." In *Heterogeneous Atmospheric Chemistry*. Edited by D. R. Schryer. Geophysical Monograph 26, American Geophysical Union, Washington D.C., pp. 99–102.

Baldwin, A. C. and D. M. Golden (1979). Heterogeneous atmospheric reactions: sulfuric acid aerosols as tropospheric sinks. *Science* **206**, 562–563.

Ban, L. L., D. Crawford, and H. Mansh (1975). Lattice resolution electron microscopy in structural studies of non-graphitizing carbons from polyvinylidene chloride. *Applied Crystall*. **8**, 415–420.

Barrie, L. A., R. M. Hoff, and S. M. Daggupaty (1981). The influence of mid-latitudinal pollution on haze in the Canadian Arctic. *Atmos. Environ*. **15**, 1407–1419.

Benner, W. H., R. Brodzinsky, and T. Novakov (1982). Oxidation of SO_2 in droplets which contain soot particles. *Atmos. Environ.* **16**, 1333–1339.

Bennett, C. F. (1968). *Human Influences on the Zoogeography of Panama Ibero–Americana 51*. University of California Press, Berkeley, California, 112 pp.

Bergstrom, R. W., T. P. Ackerman, and L. W. Richards (1982). "The optical properties of particulate elemental carbon." In *Particulate Carbon*. Edited by G. T. Wolff and R. L. Klimisch. Plenum Press, New York, pp. 43–51.

Berner, A., G. Reischl, and H. Puxbaum (1984). Size distribution of traffic derived aerosols. *Sci. Total Environ.* **36**, 299–303.

Bertine, K. K. and M. F. Mendeck (1978). Industrialization of New Haven, Connecticut, as recorded in reservoir sediments. *Environ. Sci. Technol.* **12**, 201–207.

Björseth, A. and G. Lunde (1979). Long-range transport of polycyclic aromatic hydrocarbons. *Atmos. Environ.* **13**, 45–53.

Blum, K. and H. Fissan (1984). Optical properties of carbonaceous particles from fires. *Sci. Total Environ.* **36**, 147–152.

Britton, L. G. and A. G. Clarke (1980). Heterogeneous reactions of sulphur dioxide and SO_2/NO_2, mixtures with a carbon soot aerosol. *Atmos. Environ.* **14**, 829–839.

Brosset, C. (1976). Airborne particles: black and white episodes. *Ambio* **5**, 157–163.

Brown, K. and J. Gentry (1984). Adsorption and condensation on polycyclic aromatic hydrocarbons on ultrafine soot aerosols. *Sci. Total Environ.* **36**, 225–232.

Budiansky, S. (1980). New attention for atmospheric carbon. *Environ. Sci. Technol.* **14**, 1430–1432.

Busek, B. R. and J. P. Bradley (1982). "Electron beam studies of individual natural and anthropogenic microparticles: compositions, structures and surface reactions." In *Heterogeneous Atmospheric Chemistry*, Edited by D. R. Schryer. Geophysical Monograph 26. American Geophysical Union, Washington, D.C., pp. 57–76.

Butler, J. D. and P. Crossley (1981). Reactivity of polycyclic aromatic hydrocarbons adsorbed on soot particles. *Atmos. Environ.* **15**, 91–94.

Butzer, X. (1971). *Environment and Archeology*. Aldine-Atherton, Chicago, Illinois, 703 pp.

Byrne, R., J. Michaelson, and A. Soutar (1977). Fossil charcoal as a measure of wildfire frequency in Southern California: a preliminary analysis. Presented at the Symposium on Environmental Consequences of Fire and Fuel Management in Mediterranean Ecosystems, Palo Alto, California, August 1–5.

Cadle, S. H. and P. J. Groblicki (1982). "An evaluation of methods for the determination of organic and elemental carbon in particulate samples." In *Particulate Carbon*. Edited by G. T. Wolff and R. L. Klimisch. Plenum Press, New York, pp. 89–109.

Cadle, S. H., P. J. Groblicki, and P. A. Mulawa (1983). Problems in the sampling and analysis of particulate carbon. *Atmos. Environ.* **17**, 593–600.

Cadle, S. H., P. J. Groblicki, and D. P. Stroup (1980). Automated carbon analyzer for particulate samples. *Anal. Chem.* **52**, 2201–2206.

Cass, G. R., P. M. Boone, and E. S. Macias (1982). "Emission and air quality relationships for atmospheric carbon particles in Los Angeles." In *Particulate Carbon: Atmospheric Life Cycle*. Edited by G. T. Wolff and R. L. Klimisch. Plenum Press, New York, pp. 207–243.

Cass, G. R., M. H. Conklin, J. J. Shaw, J. J. Huntzicker, and E. S. Macias (1984). Elemental carbon concentrations: estimation of an historical data base. *Atmos. Environ.* **18**, 153–162.

Chang, S. G., R. Brodzinsky, L. A. Gundel, and T. Novakov (1982). "Chemical and catalytic properties of elemental carbon." In *Particulate Carbon*. Edited by G. T. Wolff and R. L. Klimisch. Plenum Press, New York, pp. 159–181.

Chang, S. G., R. Brodzinsky, R. Toossi, S. S. Markowitz, and T. Novakov (1979). "Catalytic oxidation of SO_2 on carbon in aqueous suspension." In Carbonaceous Particles in the Atmosphere. Edited by T. Novakov. Lawrence Berkeley Laboratory, Berkeley, California, pp. 122–130.

Chang, S., J. W. Smith, I. Kaplan, J. Lawless, K. A. Kvenvolden, and C. Ponnamperuma (1970). Carbon compounds in lunar fines from Mare Tranquillitas-IV. Evidence for oxides and carbides. *Proc. Apollo 11 Lunar Sci. Conf.* **2**, 1857–1869.

Chang, S. G. and T. Novakov (1975). Formation of pollution particulate nitrogen compounds by NO—Soot and NH_3—Soot gas particle surface reactions. *Atmos. Environ.* **9**, 495–504.

Chang, S. G., R. Toossi, and T. Novakov (1981). The importance of soot particles and nitrous acid in oxidizing SO_2 in atmospheric aqueous droplets. *Atmos. Environ.* **15**, 1287–1292.

Charlson, R. J. and J. A. Ogren (1982). "The atmospheric cycle of elemental carbon." In *Particulate Carbon*. Edited by G. T. Wolff and R. L. Klimisch. Plenum Press, New York, pp. 3–18.

Cheng, R. J., V. A. Mohnen, T. T. Shen, M. Current, and J. B. Hudson (1976). Characterization of particulates from power plants. *J. Air Poll. Control Assoc.* **26**, 787–790.

Chu, L. and E. S. Macias (1983). Visibility degradation due to fine carbonaceous aerosols in St. Louis. Abstracts of 2nd International Conference on Carbonaceous Particles in the Atmosphere. Linz.

Chylek, P., V. Ramaswamy, and V. Srivastava (1984). Graphitic carbon content of aerosols, clouds and snow and its climatic implication. *Sci. Total Environ.* **36**, 117–120.

Chylek, P., V. Ramaswamy, and V. Srivastava (1983). Albedo of soot-contaminated snow. *J. Geophys. Res.* **88**, 10,837–10,843.

Clarke, A. D., R. E. Weiss, and R. J. Charlson (1984). Elemental carbon aerosols in the urban, rural, and remote-marine troposphere and in the stratosphere: inferences from light absorption data and consequences regarding radiative transfer. *Sci. Total Environ.* **36**, 97–102.

Clemenson, M., T. Novakov, and S. S. Markowitz (1980). Determination of carbon in atmospheric aerosols by deuteron analysis. *Anal. Chem.* **52**, 1758–1761.

Cofer, W. R., D. R. Schryer, and R. S. Rogowski (1980). The enhanced oxidation of SO_2 by NO_2 on carbon particulates. *Atmos. Environ.* **14**, 571–757.

Cofer, W. R., D. R. Schryer, and R. S. Rogowski (1981). The oxidation of SO_2 on carbon particles in the presence of O_3, NO_2 and N_2O. *Atmos. Environ.* **15**, 1281–1286.

Cockburn, A., R. A. Barraco, T. A. Reyman, and W. H. Peck (1975). Autopsy of an Egyptian mummy. *Science* **187**, 155–160.

Conklin, M. H., G. R. Cass, L. Chu, and E. S. Macias (1981). "Winter type carbonaceous aerosols in Los Angeles." In *Atmospheric Aerosol*, Edited by E. S. Macias and P. K. Hopke, *ACS Symposium Series* **167**, 235–250.

Cooper, J. A., L. A. Curie, and G. A. Klouda (1981). Assessment of contemporary carbon combustion source. Contribution to urban air particulate levels using C-14 measurements. *Environ. Sci. Technol.* **15**, 1045–1049.

Cope, M. J. (1979). Physical and chemical properties of coalified and charcoalified phytoclasts from some British Mesozoic sediments: an organic geochemical approach to paleobotany. Manuscript presented at the Ninth International Meeting on Organic Geochemistry. Newcastle-upon-Tyne, September 1979.

Cope, M. J. and W. G. Chaloner (1980). Fossil charcoal as evidence of past atmospheric composition. *Nature* **283**, 647–649.

Cramer, O. P. (1974). "Air quality influences." In *Environmental Effects of Forest Residues Management in the Pacific Northwest*. Edited by O. P. Cramer. USDA Forest Service General Technical Report PNW-24, pp. F1–F51.

Creager, J. S., et al. (1973). Initial Reports of the Deep-Sea Drilling Project, Vol. 19. U.S. Gov. Printing Office, Washington, D.C., 913 pp.

Cuddihy, R. G., W. C. Griffith, and R. O. McClellan (1984). Health risks from light duty diesel vehicles. *Environ. Sci. Technol.* **18**, 14A–21A.

Currie, L. A. (1982) "Contemporary particulate carbon." In *Particulate Carbon*. Edited by G. T. Wolff and R. L. Klimisch. Plenum Press, New York, pp. 245–260.

Cwynar, L. C. (1977). Recent fire history of Barron Township, Algonquin Park. *Can. J. Bot.* **55**, 1524–1538.

Cwynar, L. C. (1978). Recent history of fire and vegetation from laminated sediment of Greenleaf Lake, Algonquin Park, Ontario. *Can. J. Bot.* **56**, 10–21.

Daisey, J. M. (1980). Organic compounds in urban aerosols. New York Academy of Sciences, New York, 50–69.

Dasch, J. M. (1982). Particulate and gaseous emission from wood burning fireplaces. *Environ. Sci. Technol.* **16**, 639–645.

Delhaye, M., P. Dhamelincourt, and F. Wallart (1979). Analysis of particulates by Raman microprobe. Toxicol. Environ. Chem. Rev. **3**, 73–87.

Del Monte, M., C. Sabbioni, and O. Vittori (1981). Airborne carbon particles and marble deterioration. *Atmos. Environ.* **15**, 645–652.

Del Monte, M., C. Sabbioni, A. Ventura, and G. Zappia (1984a). Crystal growth from carbonaceous particles. *Sci. Total Environ.* **36**, 247–254.

Del Monte, M., C. Sabbioni, and G. Vittori (1984b). Urban stone sulphation and oil-fired carbonaceous particles. *Sci. Total Environ.* **36**, 369–376.

Delumyea, R. G., L. Chu, and E. S. Macias (1980). Determination of elemental carbon component of soot in ambient aerosol samples. *Atmos. Environ.* **14**, 647–652.

Delumyea, R. D. and D. Mitchell (1983). Comparison of reflectance and photoacoustic photometry for determination of elemental carbon in aerosols. *Anal. Chem.* **55**, 1996–1999.

Ehrlich, P. R., et al. (1984). Long-term biological consequences of nuclear war. *Science* **222**, 1293–1300.

Encyclopaedia Britannica (1974), Micropedia, Vol. 2, p. 749.

EPA (1979). Compilation of air pollutant emission factors. Third Edition. Supplement No. 9, PB-295 614, 106 pp.

Eatough, D. J., W. P. Green, and L. D. Hansen (1979). "Oxidation of sulfite by activated charcoal." In *Carbonaceous Particles in the Atmosphere*. Edited by T. Novakov. Lawrence Berkeley Laboratory, Berkeley, California, pp. 131–132.

Etz, E. S., J. J. Blaha, and G. J. Rosasco (1979). "Detection and identification of airborne carbonaceous matter with a Raman microprobe." In *Carbonaceous Particles in the Atmosphere*. Edited by T. Novakov. Lawrence Berkeley Laboratory, Berkeley, California, pp. 59–69.

Filion, L. (1984). A relationship between dunes, fire and climate recorded in the Holocene deposits of Quebec. *Nature* **309**, 543–546.

Fisher, G. L., C. Crisp, and T. L. Hayes (1979). "Carbonaceous particles in coal fly ash." In *Carbonaceous Particles in the Atmosphere*. Edited by T. Novakov, pp. 229–232. Lawrence Berkeley Laboratory, Berkeley, California.

Fitch, W. L. and D. H. Smith (1979). Analysis of adsorption properties and adsorbed species on commercial polymeric carbons. *Environ. Sci. Technol.* **13**, 341–346.

Foley, G. and A. van Buren (1982). Substitutes for wood. *Unasylua* **32**, 11–24.

Fritschen, L., H. Bovee, K. Buettner, R. Charlson, L. Monteith, S. Pickford, J. Murphy, and E. Darley (1970). Slash fire atmospheric pollution. U.S. Department of Agriculture Forest Research Paper ONW 97, 42 pp.

Gaffney, J. S., R. L. Tanner, and M. Phillips (1984). Separating carbonaceous aerosol source terms using thermal evolution, carbon isotopic measurements and C/N/S determinations. *Sci. Total Environ.* **36**, 53–60.

Gay, A. J., R. F. Littlejohn, and P. J. van Duin (1984). Studies of carbonaceous cenospheres from fluidised-bed combustors. *Sci. Total Environ.* **36**, 239–246.

Goldberg, E. D., V. F. Hodge, J. J. Griffin, M. Koide, and D. N. Edgington (1981). The impact of fossil fuel combustion on the sediments of Lake Michigan. *Environ. Sci. Technol.* **15**, 466–471.

Gorham, E. (1955). On the acidity and salinity of rain. *Geochim. Cosmochim. Acta* **7**, 231–239.

Gowlett, J. A. J., W. K. Harris, D. Walton, and B. A. Wood (1981). Early archaeological sites, hominid remains and track of fire from Chesowanja, Kenya. *Nature* **294**, 125–129.

Gray, H. A., G. R. Cass, J. J. Huntzicker, E. K. Heyerdahl, and J. A. Rau (1984). Elemental and organic carbon particle concentrations: a long-term perspective. *Sci. Total. Environ.* **36**, 17–25.

Griffin, J. J. and E. D. Goldberg (1975). The fluxes of elemental carbon in coastal marine sediments. *Limnol. Oceanogr.* **20**, 456–463.

Griffin, J. J. and E. D. Goldberg (1979). Morphologies and origin of elemental carbon in the environment. *Science* **206**, 563–565.

Griffin, J. J. and E. D. Goldberg (1981). Sphericity as a characteristic of solids from fossil fuel burning in Lake Michigan sediment. *Geochim. Cosmochim. Acta* **45**, 763–769.

Griffin, J. J. and E. D. Goldberg (1983). Impact of fossil fuel combustion on sediments of Lake Michigan: A reprise. *Environ. Sci. Technol.* **17**, 244–245.

Groblicki, P. J., G. T. Wolff and R. J. Countess (1981). Visibility reducing species in the Denver "Brown Cloud"—I. Relationships between extinction and chemical composition. *Atmos. Environ.* **15**, 2473–2484.

Gundel, L. A., R. L. Dod, H. Rosen, and T. Novakov (1984). The relationship between optical attenuation and black carbon concentration for ambient and source particles. *Sci. Total Environ.* **36**, 197–202.

Habenreich, T. and H. Horvath (1984). Variability of light absorption in Vienna. *Sci. Total Environ.* **36**, 141–146.

Hahn-Weinheimer, P. and A. Hirner (1981). Isotopic evidence for the origin of graphite. *Geochem. J.* **15**, 9–15.

Hansen, A. D. A. and H. Rosen (1984). Vertical distributions of particulate carbon, sulfur and bromine in the Arctic haze and comparison with ground-level measurements at Barrow, Alaska. *Geophys. Res. Lett.* **11**, 381–384.

Hansen, A. D. A., H. Rosen, and T. Novakov (1984). The aethalometer—an instrument for the real time measurement of optical absorption by aerosol particles. *Sci. Total Environ.* **36**, 191–196.

Hansen, A. D. A., H. Rosen, R. I. Dod, and T. Novakov (1979). "Optical characterization of ambient and source particles." In *Carbonaceous Particles in the Atmosphere*. Edited by T. Novakov. Lawrence Berkeley Laboratory, Berkeley, California, pp. 116–121.

Harris, S. J. and A. M. Weiner (1983a). Surface growth of soot particles in premixed ethylene air flames. *Combust. Sci. Technol.* **31**, 155–167.

Harris, S. J. and A. M. Weiner (1983b). Determination of the rate constant for soot surface growth. *Combust. Sci. Technol.* **32**, 267–275.

Harris, T. M. (1957). A Liasso-Rhaetic flora in South Wales. *Proc. R. Soc. London* **147B**, 289–308.

Harris, T. M. (1958). Forrest fire in the Mesozoic. *J. Ecol.* **46**, 447–453.

Harrison, W. E. (1976). Laboratory graphitization of a modern estuarine kerogen. *Geochim. Cosmochim. Acta* **40**, 247–248.

Heintzenberg, J. (1982). Size-segregated measurements of particulate elemental carbon and aerosol light absorption at remote arctic locations. *Atmos. Environ.* **16**, 2461–2469.

Heintzenberg, J. and P. Winkler (1984). Elemental carbon in the urban aerosol: results of a seventeen month study in Hamburg, Federal Republic of Germany, *Sci. Total Environ.* **36**, 27–38.

Heintzenberg, J. and D. S. Covert (1984). Size distribution of elemental carbon, sulfur and total mass in the radius range 10^{-6} to 10^{-4} cm. *Sci. Total Environ.* **36**, 289–297.

Heintzenberg, J. (1982). "Measurement of light absorption and elemental carbon in atmospheric aerosol samples from remote locations." In *Particulate Carbon*. Edited by G. T. Wolff and R. L. Klimisch. Plenum Press, New York, pp. 371–377.

Heisler, S. L., R. C. Henry, J. G. Watson, and G. M. Hidy (1980). The 1978 Denver winter haze study. Environmental Research and Technology Inc. Document No. P-5417-1/2, Prepared for Motor Vehicle Manufacturers Association of the United States, Inc.

Henderson, T. R., J. D. Sun, R. E. Royer, C. R. Clark, A. P. Li, T. M. Harvey, D. H. Hunt, J. E. Fulford, A. M. Lovette, and W. R. Davidson (1983). Triple-quadrupole mass spectrometry studies of nitroaromatic emissions from different diesel engines. *Environ. Sci. Technol.* **17**, 443–449.

Henderson, T. R., J. D. Sun, A. P. Li, R. L. Hanson, W. E. Bechtold, T. M. Harvey, J. Shabanowitz, and D. F. Hunt (1984). GC/MS and MS/MS studies of diesel exhaust mutagenicity and emissions from chemically defined fuels. *Environ. Sci. Technol.* **18**, 428–434.

Herring, J. R. (1977). Charcoal fluxes into Cenozoic sediments of the North Pacific: Unpublished Ph.D. thesis, University of California at San Diego, California, 105 pp.

Hock, J. L. and D. Lichtman (1982). Studies of surface layers on single particles of in-stack coal fly ash. *Environ. Sci. Technol.* **16**, 423–427.

Hopkins, B. (1965). Observations on Savanna burning in the Olokemeje Forest Reserve, Nigeria. *J. Appl. Ecol.* **2**, 367–381.

Heusser, C. J. (1983). Quaternary pollen record from Laguna de Tagua, Tagua, Chile. *Science* **219**, 1429–1431.

Hulett, L. D., A. J. Weinberger, K. J. Northcutt, and M. Ferguson (1980). Chemical species in fly ash from coal-burning power plants. *Science* **210**, 1356–1358.

Huntzicker, J. J., R. L. Johnson, J. J. Shaw, and R. A. Cary (1982). "Analysis of organic and elemental carbon in ambient aerosols by a thermal optical method." In *Particulate Carbon*. Edited by G. T. Wolff and R. L. Klimisch. Plenum Press, New York, pp. 79–88.

Issac, G. L. (1977). Olorgesailie. University of Chicago Press, Chicago, Illinois, 272 pp. p. 94.

Isono, K., M. Komabayasi, T. Takeda, T. Tanaka, K. Iwai, and M. Fujiwara (1971). Concentration and nature of ice nuclei in rim of the North Pacific Ocean. *Tellus* **23**, 40–59.

Japar, S. M., A. C. Szkarlat, and W. R. Pierson (1984). The determination of the optical properties of airborne particle emissions from diesel vehicles. *Sci. Total Environ.* **36**, 121–130.

Japar, S. M., A. C. Szkarlat, R. A. Gorse, Jr., E. K. Heyerdahl, R. L. Johnson, J. A. Rau, and J. J. Huntzicker (1984). Comparison of solvent extraction and thermal-optical carbon analysis methods: application to diesel vehicle exhaust aerosol. *Environ. Sci. Technol.* **18**, 231–234.

Jedwab, J. and J. Boulègue (1984). Graphite crystals in hydrothermal vents. *Nature* **310**, 41–43.

Jenkins, G. M., K. K. Kawamura, and L. L. Ban (1972). Formation and structure of polymeric carbons. *Proc. R. Soc. London* **327A**, 501–517.

Jensen, T. E. and R. A. Hites (1983). Aromatic diesel conditions as a function of engine conditions. *Anal. Chem.* **55**, 594–599.

Jones, E. W. (1945). The structure and reproduction of the virgin forest of the north temperate zone. *New Phytol.* **44**, 130–148.

Johnson, R. L. and J. J. Huntzicker (1979). "Analysis of volatizable and elemental carbon in ambient aerosols." In *Carbonaceous Particles in the Atmosphere*. Edited by T. Novakov. Lawrence Berkeley Laboratory, Berkeley, California, pp. 10–13.

Joranger, E. and B. Ottar (1984). Air pollution studies in the Norwegian Arctic. *Geophys. Res. Lett.* **11**, 365–368.

Judeikis, H. S., T. B. Stewart, and A. G. Wren (1978). Laboratory studies of heterogeneous reactions of SO_2. *Atmos. Environ.* **12**, 1633–1641.

Kaden, D. A. and W. G. Thilly (1979). "Mutagenic activity of fossil fuel combustion products." In *Carbonaceous Particles in the Atmosphere*. Edited by T. Novakov. Lawrence Berkeley Laboratory, Berkeley, California, pp. 193–198.

Kalb, J. E., C. J. Jolly, E. B. Oswald, and P. F. Whitehead (1984). Early hominid habitation in Ethiopia. *Am. Sci.* **72**, 168–178.

Keifer, J. R., M. Novicky, M. S. Akhter, A. R. Chughtai, and D. M. Smith (1981). The nature and reactivity of the elemental carbon (soot) as revealed by Fourier transform (FT–IR) spectroscopy. Procedures of the 1981 International Conference on Fourier Transform Infrared Spectroscopy. Vol. 289, pp. 184–188.

Klempier, N. and H. Binder (1983). Determination of polycyclic aromatic hydrocarbons in soot by mass spectrometry with direct sample insertion. *Anal. Chem.* **55**, 2104–2106.

Kneip, T. J., J. M. Daisey, J. J. Solomon, and R. J. Hershman (1983). *N*-nitroso compounds: evidence for their presence in airborne particles. *Science* **221**, 1045–1047.

Knoevenagel, K. and R. Himmelreich (1976). Degradation of compounds containing carbon atoms by photooxidation in the presence of water. *Arch. Environ. Contam. Toxicol.* **4**, 324–333.

Kothari, B. K. and M. Wahlen (1984). Concentration and surface morphology of charcoal particles in sediments of Green Lake, N.Y.: implications regarding the use of energy in the past. *Northeastern Environ. Sci.* **3**, 24–29.

Ksenzhek, O. S. and Z. V. Solevei (1960). Kinetics of the oxidation of graphite by hypochlorite and hypochlorous acid. *Zh. Prikl. Khim.* **33**, 279–283.

Kukreja, V. P. and J. L. Bove (1976). Determination of free carbon collected on a high-volume glass fiber filter. *Environ. Sci. Technol.* **10**, 187–189.

Kumai, M. (1976). Identification of nuclei and concentrations of chemical species in snow crystals sampled at the south pole. *J. Atmos. Sci.* **33**, 833–341.

Lee, M. L., D. W. Later, D. K. Rollins, D. J. Eatough, and L. D. Hansen (1980). Dimethyl and monomethyl sulfate: presence in coal fly ash and airborne particulate matter. *Science* **207**, 186–188.

Lipfert, F. W. and J. L. Dungan (1983). Residential firewood use in the United States. *Science* **219**, 1425–1427.

Long, J. R. and D. J. Hansen (1983). Dioxin issue focuses on three major controversies in the U.S. *Chem. Eng. News* **61**, 23–36.

Lunde, G. and A. Björseth (1977). Polycyclic aromatic hydrocarbons in long-range transported aerosols. *Nature* **268**, 518–519.

Macias, E. S., C. D. Radcliffe, C. W. Lewis, and C. R. Sawicki (1978). Proton-induced x-ray analysis of atmospheric aerosols for carbon, nitrogen and sulfur composition. *Anal. Chem.* **50**, 1120–1124.

Macias, E. S. and L. C. Chu (1982). "Carbon analysis of atmospheric aerosols using GRALE and Reflectance Analysis." In *Particulate Carbon*. Edited by G. T. Wolff and R. L. Klimisch. Plenum Press, New York, pp. 131–144.

Macias, E. S., R. Delumyea, L. Chu, H. R. Appleman, C. D. Radcliffe, and L. Stanley (1979). "The determination, speciation and behavior of particulate carbon." In *Carbonaceous Particles in the Atmosphere*. Edited by T. Novakov. Lawrence Berkeley Laboratory, Berkely, California, pp. 70–83.

Mantell, C. L. (1968), *Carbon and Graphite Handbook*. Interscience, New York, 538 pp.

Mathez, E. A. and J. R. Delaney (1981). The nature and distribution of carbon in submarine basalts and peridotite nodules. *Earth Planet. Sci. Lett.* **56**, 217–232.

Matteson, M. J. (1979). "Capture of atmospheric gases by water vapor condensation on carbonaceous particles." In *Carbonaceous Particles in the Atmosphere*. Edited by T. Novakov. Lawrence Berkeley Laboratory, Berkeley, California, pp. 150–154.

Mattson, J. S. and H. B. Mark, Jr. (1971). *Activated Carbon*. Marcel Dekker, New York, 237 pp.

McElroy, M. W., R. C. Carr, D. S. Ensor, and G. R. Markowski (1982). Size distribution of fine particles from coal combustion. *Science* **215**, 13–19.

McGraw-Hill Encyclopedia of Science and Technology, Vol. 3, p. 16 (1977).

Medalia, A. I. and F. A. Heckman (1969). Morphology of aggregates—II. Size and shape factors of carbon black aggregates from electron microscopy. *Carbon* **7**, 567–582.

Medalia, A. I. (1974). Filter aggregates and their effect on reinforcement. *Rubber Chem. Technol.* **47**, 411–433.

Medalia, A. I. and D. Rivin (1981). Forms of particulate carbon in soot and carbon black. Extended Abstracts, Program. Biennial Conference Carbon, 15th, pp. 480–481.

Medalia, A. I. and D. Rivin (1982). Particulate carbon and other components of soot and carbon black. *Carbon* **20**, 481–492.

Meszaros, A. (1984). The number, concentration and size distribution of the soot particles in the 0.02–0.5 micron radius range at sites of different pollution levels. *Sci. Total Environ.* **36**, 283–288.

Michel-Levy, M. C. and A. Lautie (1981). Microanalysis by Raman spectroscopy of carbon in the Tieschitz chondrite. *Nature* **292**, 321–322.

Middleton, P., C. S. Kiang, and V. A. Mohnen (1982). "The relative importance of various urban sulfate aerosol production mechanisms—a theoretical comparison." In *Heterogenous Atmospheric Chemistry*, Edited by D. R. Schryer. Geophysical Monograph 26. American Geophysical Union, Washington, D.C., pp. 221–230.

Minerals Yearbook (1976 and 1982). U.S. Government Printing Office, Vol. 1.

Minnich, R. A. (1983). Fire mosaics in Southern California and Northern Baja California. *Science* **219**, 1287–1294.

Mueller, P. K., R. W. Mosley, and L. B. Pierce (1971). "Carbonate and noncarbonate carbon in atmospheric particles." In *Proceedings of the 2nd International Clean Air Congress*, Edited by H. M. Englund and W. T. Beery. Academic Press, New York, pp. 532–539.

Mueller, P. K., K. K. Fung, S. L. Heisler, D. Grosjean, and G. M. Hidy (1982).

"Atmospheric particulate carbon observations in urban and rural areas of the United States." In *Particulate Carbon*. Edited by G. T. Wolff and R. L. Klimisch. Plenum Press, New York, pp. 343–370.

Muhlbaier, J. L. and R. L. Williams (1982). "Fireplaces, furnaces and vehicles as emission sources of particulate carbon." In *Particulate Carbon*. Edited by G. Wolff and R. L. Klimisch. Plenum Press, New York, pp. 185–205.

Muller, J. (1984). Atmospheric residence time of carbonaceous particles and particulate PAH-compounds. *Sci. Total Environ.* **36**, 339–346.

Murphy, J. L., L. J. Fritschzen, and O. P. Cramer (1970). Research looks at air quality and forest burning. *J. For.* **68**, 530–535.

NAS (1981). *NO_x Emission Controls for Heavy-Duty Vehicles: Toward Meeting a 1986-Standard*. National Academy Press, Washington, D.C., 127 pp.

Natusch, D. F. S. (1976). "Characterization of atmospheric pollutants from power plants." In *Proceedings of the Second Federal Conference on the Great Lakes, 1975*. Great Lakes Basin Commission, Ann Arbor, Michigan.

Novakov, T., S. G. Chang, and A. B. Harker (1974). Sulphates as pollution particulates: catalytic formation on carbon (soot) particles. *Science* **186**, 259–261.

Novakov, T. (1982). "Soot catalyzed atmospheric reactions." In *Heterogeneous Atmospheric Chemistry*. Edited by D. R. Schryer. Geophysical Monograph 26. American Geophysical Union, Washington, D.C., pp. 215–220.

Novakov, T. (1984). The role of soot and primary oxidants in atmospheric chemistry. *Sci. Total Environ.* **36**, 1–10.

Novakov, T., P. K. Mueller, A. E. Alcocer and J. V. Otvos (1972). Chemical composition of Pasadena aerosol by particle size and time of day III. Chemical states of nitrogen and sulfur by photoelectron spectroscopy. *J. Colloid. Interfac. Sci.* **39**, 225–234.

NRC (1982). Diesel Cars. Benefits, Risk and Public Policy. National Academy Press, Washington, D.C., 142 pp.

Ogren, John (1982). "Deposition of particulate elemental carbon from the atmosphere." In *Particulate Carbon*. Edited by G. T. Wolff and R. L. Klimisch. Plenum Press, New York, pp. 379–391.

Ogren, J. A. and R. J. Charlson (1983). Elemental carbon in the atmosphere: cycle and lifetime. *Tellus* **35B**, 241–254.

Ogren, John, R. J. Charlson, and P. J. Groblicki (1983). Determination of elemental carbon in rainwater. *Anal. Chem.* **55**, 1569–1572.

Ogren, J. A., P. J. Groblicki, and R. J. Charlson (1984). Measurement of the removal rate of elemental carbon from the atmosphere. *Sci. Total Environ.* **36**, 329–338.

Parkin, D. W., D. R. Phillips, and R. A. L. Sullivan (1970). Airborne dust collections, over the North Atlantic. *J. Geophys. Res.* **75**, 1782–1793.

Patterson, E. M. (1979). "optical properties of urban aerosols containing carbonaceous material." In *Carbonaceous Particles in the Atmosphere*. Edited by

T. Novakov. Lawrence Berkeley Laboratory, Berkeley, California, pp. 247–251.

Peaden, P. A., M. L. Lee, Y. Hirata, and M. Novotny (1980). High-performance liquid chromatographic separation of high molecular weight polycyclic aromatic compounds in carbon black. *Anal. Chem.* **52**, 2268–2271.

Pheiffer, J. E. (1972). *The Emergence of Man*. Harper and Row, New York, p. 165.

Pierce, R. C. and M. Katz (1975). Dependence of polynuclear aromatic hydrocarbon content on size distribution of atmospheric aerosols. *Environ. Sci. Technol.* **9**, 347–353.

Pierce, R. C. and M. Katz (1976). Chromatographic isolation and spectral analysis of polycyclic quinones. Application to air pollution analysis. *Environ. Sci. Technol.* **10**, 45–51.

Pierson, W. R. (1979), "Particulate organic matter and total carbon from vehicles on the road." In *Carbonaceous Particles in the Atmosphere*. Edited by T. Novakov. Lawrence Berkeley Laboratory, Berkely, California, pp. 221–228.

Potter, M. C. (1908). Bacteria as agents in the oxidation of amorphous carbon. *Proc. R. Soc. London* **80B**, 239–259.

Puxbaum, H. and H. Baumann (1984). Vertical concentration profiles of traffic derived components in a street canyon. *Sci. Total Environ.* **36**, 47–52.

Pyne, S. J. (1982). *Fire in America*. Princeton University Press, Princeton, New Jersey.

Rahn, K. A. (1981a). Relative importances of North America and Eurasia as sources of Arctic aerosol. *Atmos. Environ.* **8**, 1447–1455.

Rahn, K. A. (1981b). The Mn/V ratio as a tracer of large scale sources of pollution aerosol for the Arctic. *Atmos. Environ.* **15**, 1457–1464.

Rahn, H. A. and D. H. Lowenthal (1984). Elemental tracers of distant regional pollution aerosols. *Science* **223**, 132–139.

Rahn, K. A., C. Brosset, B. Ottar, and E. M. Patterson (1982). "Black and white episodes, chemical evolution of Eurasian air masses and long range transport of carbon to the Arctic." *Particulate Carbon*. Edited by G. T. Wolff and R. L. Klimisch. Plenum Press, New York, pp. 327–342.

Rahn, K. A. and R. J. McCaffrey (1980). On the origin and transport of the winter Arctic aerosol. *Ann. N.Y. Acad. Sci.* **338**, 486.

Ramdahl, T., J. Schjoldager, L. A. Currie, J. E. Hanssen, M. Moller, G. A. Klouda, and I. Alfheim (1984). Ambient impact of residential wood combustion in Elverum, Norway. *Sci. Total Environ.* **36**, 81–90.

Reif, A. E. (1981). The causes of cancer. *Am. Sci.* **69**, 437–447.

Roessler, D. M. (1984). Photoacoustic insights on diesel exhaust particles. *Sci. Total Environ.* **36**, 183–190.

Roessler, D. M. and F. R. Faxvog (1980). Optical properties of agglomerated acetylene smoke particles at 0.5145 μm and 10.6 μm wavelengths. *J. Opt. Soc. Am.* **70**, 230–235.

Rogowski, R. S., D. R. Schryer, W. R. Coffer, III, and R. A. Edahl, Jr. (1982).

"Oxidation of SO_2 by NO_2 and air in aqueous suspension of carbon." In *Heterogeneous Atmospheric Chemistry*. Edited by D. R. Schryer. Geophysical Monograph 26, Washington, D.C., pp. 174–177.

Rosen, H. and T. Novakov (1977). Raman scattering and the characterization of atmospheric aerosol particles. *Nature* **266**, 708–710.

Rosen, H., A. D. A. Hansen, R. L. Dod, L. A. Gundel, and T. Novakov (1982). "Graphitic carbon in urban environments and the Arctic (1982)." In *Particulate Carbon*. Edited by G. T. Wolff and R. L. Klimish. Plenum Press, New York, pp. 273–293.

Rosen, H., T. Novakov, and B. A. Bodhaine (1981). Soot in the Arctic. *Atmos. Environ*. **15**, 1371–1374.

Rosen, H., A. D. A. Hansen, L. Gundel, and T. Novakov (1978). Identification of the optically absorbing component in urban aerosols. *Appl. Opt*. **17**, 3859–3861.

Rosen, H., A. D. A. Hansen, R. L. Dod, L. A. Gundel, and T. Novakov (1982). "Graphitic carbon in urban environments and the Arctic." In *Particulate Carbon*. Edited by G. T. Wolff and R. L. Klimisch. Plenum Press, New York, pp. 273–293.

Rosen, H., A. D. A. Hansen, L. Gundel, and T. Novakov (1979). "Identification of the graphitic carbon component of source and ambient particulates by Raman spectroscopy and an optical attenuation technique." In *Proceedings of the Conference on Carbonaceous Particles in the Atmosphere*. Edited by T. Novakov. Report LBL-9037. Lawrence Berkeley Laboratory, Berkeley, California, 1979, pp. 49–55.

Rosen, H., A. D. A. Hansen, and T. Novakov (1984). Role of graphitic carbon particles in radiative transfer in the Arctic haze. *Sci. Total Environ*. **36**, 103–110.

Rosen, H. and A. D. A. Hansen (1984). Role of combustion-generated carbon particles in the absorption of solar radiation in the Arctic haze. *Geophys. Res. Lett*. **11**, 461–464.

Rosen, H., A. D. A. Hansen, R. L. Dod, and T. Novakov (1980). Soot in urban atmospheres: determination by an optical absorption technique. *Science* **208**, 741–744.

Rotty, R. M. (1981). "Data for global CO_2 production from fossil fuels and cement." In *Carbon Cycle Modelling*. Edited by B. Bolin. SCOPE 16. John Wiley & Sons, New York, pp. 121–125.

Russell, P. A. (1979). "Carbonaceous particulates in the atmosphere: illumination by electron microscopy." In *Carbonaceous Particles in the Atmosphere*. Edited by T. Novakov. Lawrence Berkeley Laboratory, Berkeley, California, pp. 133–140.

Sadler, M., R. J. Carlson, H. Rosen, and T. Novakov (1981). An intercomparison of the integrating plate and the laser transmission methods for determination of aerosol absorption measurements. *Atmos. Environ*. **15**, 1265–1268.

Sanak, J., A. Gaudry, and G. Lambert (1981). Size distribution of Pb-210 aerosol over oceans. *Geophys. Res. Lett.* **8**, 1067–1069.

Saltzman, E. S., G. W. Brass, and D. A. Price (1983). The mechanism of sulfate aerosol formation: chemical and sulfur isotope evidence. *Geophys. Res. Lett.* **10**, 513–516.

Sandberg, D. V., S. G. Pickford, and E. F. Darley (1975). Emissions from slash burning and the influence of flame retardent chemicals. *J. Air Poll. Control Assoc.* **25**, 278–281.

Sandberg, D. V. (1974). Slash fire intensity and smoke emissions. Presented at Third National Conference on Fire and Forest Meteorology of the American Meteorological Society and the Society of American Forestry, April 2–4 (1974). Lake Tahoe, California.

Sanford, R. L. Jr., J. Saldarriaga, K. E. Clark, C. Uhl, and R. Herrera (1985). Amazon rain forest fires. *Science* **227**, 53–55.

Schnell, R. C. and W. E. Raatz (1984). Vertical and horizontal characteristics of Arctic haze during AGASP: Alaskan Arctic. *Geophys. Res. Lett.* **11**, 369–372.

Schopf, J. M. (1975). Modes of fossil preservation. *Rev. Palaeobot. Palynol.* **20**, 27–53.

Schryer, D. R., W. R. Cofer, III, and R. S. Rogowski (1980). Synergistic effects in trace gases—aerosol interactions. *Science* **209**, 723.

Schryer, D. R., R. S. Rogowski, and W. R. Cofer, III (1982). Soot-catalyzed reactions. *Science* **216**, 1174.

Scott, A. (1974). The earliest conifer. *Nature* **251**, 707–708.

Scott, A. C. and M. E. Collinson (1978). "Organic sedimentary particles: results from scanning electron microscope studies of fragmentary plant material." In *Scanning Electron Microscopy in the Study of Sediments*. Edited by W. Brian Whalley. Geo Abstracts, Norwich, England, pp. 137–167.

Seiler, W. and P. J. Crutzen (1980). Estimates of gross and net fluxes of carbon between the biosphere and the atmosphere from biomass burning. *Climatic Change* **2**, 207–247.

Shneour, E. (1966). Oxidation of graphite carbon in certain soils. *Science* **151**, 991–992.

Simmons, I. G. and J. B. Innes (1981). Tree remains in a North York Moors spat profile. *Nature* **294**, 76–78.

Skolnick, H. (1958). Stratigraphy of some lower Cretaceous rocks of Black Hills area. *Bull. Am. Assoc. Pet. Geol.* **42**, 787–815.

Slatkin, D. N., L. Friedman, A. P. Irsa, and J. S. Gaffney (1978). The C-13/C-12 ratio in black pulmonary pigment: a mass spectrometric study. *Hum. Pathol.* **9**, 259–267.

Smith, D. M., J. J. Griffin, and E. D. Goldberg (1973). Elemental carbon in marine sediments: a baseline for burning. *Nature* **241**, 268–270.

Smith, D. M., J. J. Griffin, and E. D. Goldberg (1975). Spectrometric method for the quantitative determination of elemental carbon. *Anal. Chem.* **47**, 233–238.

Smith, P. P. K. and P. R. Buseck (1981). Graphitic carbon in the Allende Meteorite. *Science* **212**, 322–324.

Smith, P. P. K. and P. R. Buseck (1982). Carbon in the Allende meteorite: evidence for graphite rather than carbyne. *Geochem. Cosmochim. Acta Suppl.* **16**, 1165–1175.

Smith, W. R. (1964). *Encyclopedia of Chemical Technology*, Vol. 4. Interscience, New York, pp. 243–280.

Spengler, J. D. and K. Sexton (1983). Indoor air pollution: a public health perspective. *Science* **221**, 9–17.

Stevens, R. K., T. G. Dzubay, R. W. Shaw, Jr., W. A. McClenny, C. W. Lewis, and W. E. Wilson (1980). Characterization of the aerosol in the Great Smoky Mountains. *Environ. Sci. Technol.* **12**, 1491–1498.

Stewart, O. C. (1956). "Fire as the first great force employed by man." In *Man's Role in Changing the Face of the Earth*. Edited by William L. Thomas, Jr. University of Chicago Press, Chicago, Illinois, pp. 115–133.

Stewart, R. B. and W. Robertson III (1971). Moisture and seed carbonization. *Econ. Bot.* **25**, 381.

Suman, D. (1983). Agricultural burning in Panama and Central America. Unpublished pH.D. Thesis, University of California at San Diego, 156 pp.

Swain, A. M. (1973). A history of fire and vegetation in Northeastern Minnesota as recorded in lake sediments. *Quat. Res.* **3**, 383–396.

Tanner, R. L., J. S. Gaffney, and M. F. Phillips (1982). Determination of organic and elemental carbon in atmospheric aerosol samples by thermal evolution. *Anal. Chem.* **54**, 1627–1630.

Thomas, J. F., M. Mukai, and B. D. Tebbens, (1968). Fate of airborne Benzo(a)pyrene. *Environ. Sci. Technol.* **2**, 33–39.

Thompson, B. *The Complete Works of Count Rumford*, American Academy of Sciences, 1870–1873. Vol. II, pp. 542–544.

Thompson, R., J. Bloemendal, J. A. Dearing, F. Oldfield, T. A. Rummery, J. C. Stober, and G. M. Turner (1980). Environmental applications of environmental measurements. *Science* **207**, 481–486.

Toon, O. W. and J. B. Pollack (1980). Atmospheric aerosols and climate. *Am. Sci.* **68**, 268–277.

Trijonis, J. (1984). Effect of diesel vehicles on visibility in California. *Sci. Total Environ.* **36**, 131–140.

Turco, R. P., O. B. Toon, T. P. Ackerman, J. B. Pollack, and C. Sagan (1984). Nuclear winter: global consequences of multiple nuclear explosions. *Science* **222**, 1283–1292.

Van Vaeck, L., G. Broddin, and K. van Cauwenberghe (1979). Differences in particle size distributions of major organic pollutants in ambient aerosols in urban, rural and seashore areas. *Environ. Sci. Technol.* **13**, 1494–1502.

Waddinton, J. C. B. (1969). A stratigraphic record of the pollen influx to a lake in the Big Woods of Minnesota. *Geol. Soc. Am. Spec. Pap.* **123**, 263–282.

Wagner, H. G. G. (1978). Soot formation in combustion. Seventeenth Symposium on Combustion. The Combustion Institute, Pittsburgh, PA, pp. 3–19.

Walsh, J. E. (1984). Snow cover and atmospheric variability. *Am. Sci.* **72**, 50–57.

Warner, P. O. (1976). Analysis of Air Pollutants. John Wiley & Sons, New York, p. 5.

Warren, S. G. and W. J. Wiscombe (1980). A model for the spectral albedo of snow II. Snow containing atmospheric aerosols. *J. Atmos. Sci.* **37**, 2734–2745.

Weiss, P., I. Friedman, and J. D. Gleason (1981). The origin of epigenetic graphite. *Geochim. Cosmochim. Acta* **45**, 2325–2332.

Weiss, R. E. and A. P. Waggoner (1982). "Optical measurements of airborne soot in urban, rural and remote locations." In *Particulate Carbon*. Edited by G. T. Wolff and R. L. Klimisch. Plenum Press, New York, pp. 317–324.

Whitby, K. T. (1979). "Size distribution and physical properties of combustion aerosols." In *Carbonaceous Particles in the Atmosphere*. Edited by T. Novakov. Lawrence Berkeley Laboratory, University of California, Berkeley, California, pp. 201–208.

White, G. F. (1972). History of fire in North America. Fire in the Environment Symposium Proceedings May 1–5, 1972, Denver, Colorado.

Wolff, G. T., R. J. Countess, P. J. Groblicki, M. A. Ferman, S. H. Cadle, and J. L. Muhlbaier (1980). Visibility reducing species in the Denver "Brown Cloud." Part II. Sources and temporal patterns; General Motors Research Laboratories Publication GMR-3394, ENV-80, 77 pp.

Wolff, G. T., N. A. Kelly, M. A. Ferman, and M. L. Morrissey (1983). Rural measurements of the chemical composition of air borne particles in the Eastern United States. *J. Geophys. Res.* **88**, 10769–10775.

Wolff, G. T., P. J. Groblocki, S. H. Cadle, and R. J. Countess (1982). "Particulate carbon at various locations in the United States." In *Particulate Carbon*. Edited by G. T. Wolff and R. L. Klimisch. Plenum Press, New York, pp. 297–315.

Yang, R. T. and C. Wong (1981). Mechanism of single-layer graphite oxidation: evaluation by electron microscopy. *Science* **214**, 437–438.

Yasa, Z., N. M. Amer, H. Rosen, A. D. A. Hansen, and T. Novakov (1979). Photoacoustic investigation of carbon aerosol particles. *Appl. Opt.* **18**, 2528–2530.

Yasuhara, A., M. Morita, and K. Fuwa (1982). Determination of naphtho(2,1,8-9qra) naphthacene in soots. *Environ. Sci. Technol* **16**, 805–808.

Yu, Ming-Li and R. A. Hites (1981). Identification of organic compounds on diesel engine soot. *Anal. Chem.* **53**, 951–954.

[illegible] major [illegible] phenolic [illegible]
[illegible]
Wa[illegible] [illegible]
[illegible]
[illegible]
[illegible]

[illegible]
[illegible]

[illegible]
[illegible]

[illegible]
[illegible]

[illegible]
[illegible]

[illegible]
[illegible]

[illegible]
[illegible]

[illegible]
[illegible]

[illegible] Science [illegible]

[illegible]
[illegible]

[illegible]
[illegible]

INDEX